RICHARD SPRUCE

(1817-1893)

BOTANIST AND EXPLORER

RICHARD SPRUCE

(1817-1893)

BOTANIST AND EXPLORER

Edited by

M.R.D. Seaward and S.M.D. FitzGerald

ROYAL
BOTANIC
GARDENS
KEW

Published by The Royal Botanic Gardens, Kew

Production Editor: S: Dickerson

Typeset at the Royal Botanic Gardens, Kew, by
Christine Beard, Dominica Costello and Margaret Newman

Book design by Jeff Eden for Media Resources,
Royal Botanic Gardens, Kew

ISBN 0 947643 94 X

Printed and bound in Great Britain by
Whitstable Litho Ltd., Whitstable, Kent.

Contents

Dedication

This volume is dedicated to Professor Richard Evans Schultes of Harvard University in acknowledgement of his outstanding contribution to Amazonian botany and his recognition of Richard Spruce's many and varied achievements.

Preface

The Richard Spruce Conference, here recorded in this volume, was by common consent of those present a remarkable experience. The diversity of papers presented and continuing research reflects not only the spirit and genius of Richard Spruce, courageous explorer of the Amazon and meticulous observer of plants and people, but also the rich diversity of the Amazon itself.

It is with great pleasure therefore that I commend this volume to its readers, and acknowledge the hard work that went into its making. The Richard Spruce Centennial Conference was sponsored by the Linnean Society and the Royal Botanic Gardens, Kew, and organised on their behalf by Professor Mark Seaward of Bradford University. The papers were edited by Professor Seaward and Sylvia FitzGerald of the Royal Botanic Gardens, Kew.

Professor Sir Ghillean Prance
Director
Royal Botanic Gardens, Kew

Introduction

Mark R. D. Seaward

Introduction

Mark R. D. Seaward
University of Bradford, UK

It is all too true that 'a prophet is not without honour save in his own country': it is a sad and curious fact that outside a limited circle of bryologists and Kew taxonomists, Britain has paid little attention to Spruce, and it is largely due to a North American, Richard Schultes, that international awareness of Spruce's major contributions to botany has been rekindled, and numerous botanists and explorers influenced. Even in his native Yorkshire, where the study of natural history has continued to flourish mainly through the activities of the Yorkshire Naturalists' Union, of which Spruce was a member, he has been virtually forgotten; only a few enthusiasts, particularly the late W. Arthur Sledge, have championed Spruce's claim to an honoured place in botanical history.

It is to Alfred Russel Wallace, the great evolutionist, who also researched extensively in South America and thus came to know Spruce,[1] that we owe so much of our knowledge of him. After Spruce's death, Wallace compiled *Notes of a botanist on the Amazon and Andes*[2] from Spruce's notebooks and letters; he added a 27-page biographical introduction and a list of 54 books and papers published by Spruce. Volume 1 of the edited *Notes* provides a fairly continuous narrative account of Spruce's travels, but the second volume is much more disjointed, due to greater reliance upon *verbatim* quotations from letters, often at considerable length, punctuated by Wallace's annotations intended to improve the continuity. It has to be noted that even though the work ran to two volumes, those notes and letters examined were condensed by their editor; furthermore, other extensive material, particularly correspondence, was never studied in detail.[3]

Richard Schultes, who took up where Wallace left off, has added very considerably to such biographical information as was formerly available.[4] His re-examination of manuscript material has revealed that editorial condensation has on occasion blurred Spruce's intended meaning and he has also unearthed correspondence unknown to Wallace which sheds new light on Spruce's thinking, particularly in respect of evolutionary theory. Further studies of Spruce's correspondence have revealed other aspects of him, particularly his ethnobotanical interests and the sound practicality of much of his work and many of his ideas.

The following account of Richard Spruce's life and achievements relies heavily on Wallace's pioneer work and on Richard Schultes' many years of research. As he would be the first to acknowledge, much work still remains to be done. It is already clear from the present author's bibliographical researches that Wallace's statement in the *Notes*[5] that his list of Spruce's published works 'is quite complete' is by no means accurate and a fuller bibliography, together with a bibliography of publications on him, are provided in Chapter 27.[6] A listing of Spruce's surviving manuscripts and correspondence, both published and unpublished, with their locations, is given in Chapter 26.[7]

Richard Spruce was born at Ganthorpe, near Castle Howard, on 10 September 1817, and spent most of his life, apart from his travels, in the North

Fig. 1. Richard Spruce as a young man.

Yorkshire villages of Welburn and Coneysthorpe.[8] He appears to have been educated wholly by his father, and followed him into the teaching profession. From an early age, he was interested in natural history. At 16, he listed the plants of Ganthorpe, and at 19 he had written a 'Flora of the Malton District', with 485 species of flowering plants. In 1839, he became a teacher of mathematics at the Collegiate School, York, where he remained until it closed down in 1844. He devoted his time thereafter to botanical pursuits.

He rapidly acquired considerable expertise in both flowering plants and bryophytes, adding *Carex appropinquata* Schumacher and *Myrinia pulvinata* (Wahlenb.) Schimp. respectively to the British flora. According to his Diary, now held by the Linnean Society, his botanical excursions to localities throughout Yorkshire were undertaken almost weekly during the period 1841 to 1844. In 1843, Spruce made numerous visits to Teesdale, one of the most botanically rich sites in the north of England; work there provided the basis for an important paper on its bryophytes.[9] June–July 1844 was particularly notable for his four-week visit to the west of Ireland; there he stayed with Thomas Taylor, one of the world's leading bryologists, and although ill for most of the time,[10] he visited numerous sites in Kerry and Cork. His *List of the Musci and Hepaticae of Yorkshire*[11] and other published papers added innumerable records to local lists and many to the English flora. With his contacts with leading botanists of the day, he built up not only a sizeable herbarium but also a sizeable reputation. Through the help and advice of William Borrer, Sir William Hooker and George Bentham, Spruce undertook a collecting excursion to the Pyrenees from May 1845 to April 1846. The sale of sets of flowering plants was intended to defray expenses, but no doubt he found the challenge of adding to a bryophyte flora of only 169 species compiled by a leading French bryologist irresistible. Indeed, he amassed a superb collection of bryophytes, 17 of the 478 species he found being new to science; his findings were the subject of another influential paper.[12]

It was remarkable that despite his delicate health, Spruce should not only undertake a twelve-month excursion to the Pyrenees, but also allow himself to be persuaded, albeit willingly, to embark on a major long-term expedition to South America. It is unclear when he determined to explore distant lands. According to Stabler,[13] once he had worked up and disposed of his Pyreneean plants, he prepared himself for future work by studying exotic plants at Kew and at the British Museum (Natural History). It would appear that his interest in the South American flora pre-dated his selection by Sir William Hooker, Director of the Royal Botanic Gardens, Kew, to undertake botanical exploration in the Amazon valley; whether he had long wished to study the flora at first hand is only likely to be determined by studying his voluminous correspondence with what was by this time a very wide circle of botanists. However, in 1849, believing that his constant bronchial trouble might be tuberculosis and wanting above all else to see and collect in a tropical rain forest before it was too late, he decided to go to the Amazon. Hooker agreed to sell dried plant specimens collected by Spruce to herbaria in Europe and the United Kingdom to finance his trip.

Spruce sailed for South America in June 1849. His task was to investigate the flora of the Amazon valley and send back collections of herbarium specimens to

Fig. 2. Richard Spruce, a photograph taken upon his return from 15 years in the Amazon and Andes of South America.

Kew Gardens. His journey took him up many of the Amazon's tributaries, and from 1855 until his return to England in May 1864 he worked the head waters of the Amazon in the northern Andes of Peru and Ecuador. In addition to the many thousands of specimens of angiosperms which he sent back to England,[14] he also made copious collections of ferns, mosses, liverworts and lichens. Of all areas, it was the Rio Negro that most enthralled him, and where a great number of his species and genera unknown to science were collected. He calculated that between 50,000 and 80,000 angiosperm species existed in the Amazon forests; today we estimate that the Amazon flora has some 80,000 species. Few botanists have worked in the Rio Negro region, Ghillean Prance and Richard Schultes amongst them;[15] it is still a storehouse of undescribed species, and still not an easy region, even for modern botanists with all their equipment and amenities. One wonders how a frail man like Spruce could have carried on such a long and productive programme without such facilities.

After a few months spent in the region around the mouth of the Amazon, he started up-river to Santarém at the month of the Tapajos, where he met Alfred Russel Wallace.[16] Spruce explored the River Trombetas almost to the borders of British Guiana, reaching Manaus at the mouth of the Rio Negro about the end of 1850. The following year he explored the forests around Manaus, after which he set out by boat with six companions for the upper reaches of the Rio Negro; here he spent the next three years, crossing to the Orinoco by the Casiquiari and penetrating some distance into Venezuela. Spruce returned to Manaus at the close of 1854; ascending the Amazon by steamer to Peru, he proceeded by canoe to Tarapoto at the eastern foot of the Andes. Here he stayed two years, amassing a very large collection of plants, including 250 different fern species collected within a radius of a mere 25 miles. In 1857 he again descended the Amazon, travelling by canoe up the Postasa to Cañelos in Ecuador, a 14-week journey of some 500 miles. The last part of this journey through the forests of Baños at the foot of the Tunguragua volcano was the most arduous and perilous of all his journeys and much of his baggage was lost or had to be abandoned in the swollen torrents and cataracts of the Tope river. Six months later he moved on to Ambato which he made his headquarters for two years and where, in spite of the civil war then raging, he explored the region at the foot of Chimborazo, making visits to Quito and other highland areas. The period spent in the foothills of the Andes, especially at Tarapoto and in the Baños forest, yielded the richest haul of bryophytes of all his journeys.

Although no economic strings were attached to his mission, Spruce's achievements had some far-reaching economic importance.[17] He laid the botanical foundations for an understanding of the genus *Hevea*, source today of most of our natural rubber. After reaching Peru he was commissioned by the India Office to locate and collect seeds and young plants of *Cinchona*, the source of the anti-malarial medicine quinine, and send these back to Kew. On the western slopes of Mt. Chimborazo, Ecuador, in the face of formidable difficulties, he succeeded in procuring 100,000 seeds and 600 seedling plants which were successfully shipped to England.[18] It was from these shipments that the *Cinchona* plantations and industry of south-eastern Asia were developed – an economic venture which although contributing substantially to the wealth of

Fig. 3. Richard Spruce's cottage, with, from left to right, Professor Richard Schultes, Mr Winston Spruce, Professor Jack Hawkes (President of the Linnean Society) and Mrs F. Cross (present occupant of the cottage).

the countries concerned, yielded Spruce only a modest annual pension of £100 on his return home.

By now Spruce's health was seriously and permanently undermined, and to add to his physical plight he lost practically all his savings when an Ecuadorian establishment (Gutierrez & Co., Guayaquil) in which they were deposited went bankrupt in 1861. Nevertheless he continued to explore the Pacific coastal areas of Ecuador and Peru[19] for a further three years before departing for England, arriving at Southampton on 27 May 1864 after an absence of fifteen years all but a few days (Figure 2).

Spruce's South American collections of some 7,000 flowering plant species, with the exception of the palms,[20] were worked on by Bentham and others. His palms (supplemented by copious notes on living plants assembled during his travels), together with a major loan of material from the Kew Herbarium arranged by Hooker, were the subject of a detailed investigation by Spruce, resulting in the publication of an important monograph[21] in which he enumerated and characterised 118 species, more than half of which, for the most part discovered by himself, are fully described as new. His ferns, mosses, lichens and fungi were named by the leading experts of the day, while his favourite group, the liverworts,[22] he worked out for himself, first at Welburn and subsequently at Coneysthorpe where, now a chronic invalid, he spent the remainder of his life. His monumental *Hepaticae Amazonicae et Andinae*[23] describing over 700 species, nearly 500 collected by himself and over 400 of these new to science[24], remains to this day the greatest work on South American bryology.[25]

Spruce's South American moss collections were sent to William Mitten in 1863, but he was dismayed, on his return to England, to find the condition they were in, and remarked in a letter to William Wilson that

> *"Mitten ... has had them in hand nearly two years and a half, and has not yet made up a single set or found a subscriber... he has taken the best specimens of everything for himself, and has worked them up in his own herbarium so as to shine on them, no doubt, some day ... he has squashed all my mss. names, and substituted others of his own ..."*[26]

However, in the spring of 1886, Spruce distributed sets of mosses he had prepared himself. The sets first sent out were without names but with Mitten's serial numbers; these vary considerably: according to Sayre,[27] the list shows 1462 consecutive numbers with scattered addenda to 1518, and some specimens have duplicates. Later sets have Mitten's determinations which Spruce had published as a catalogue in 1867. The material provided an indispensable source for Mitten's *Musci Austro-americani* published in 1869.[28]

No botanist studying tropical South American plants and their uses can afford not to consult Spruce's specimens and detailed notes.[29] Their value for those working in the fields of phytochemistry, ethnotoxicity, hallucinogens, narcotics and economic botany is immense,[30] and his herbarium specimens are often cited in taxonomic investigations, frequently as the type material.[31] In many cases, Spruce not only first discovered a particular plant, but also recognised its economic potential through careful observation of native usage[32] and custom, often communicating with local inhabitants in their own tongues, and making the prodigious effort to learn and record for the first time vocabularies of twenty-one different Amazonian dialects.[33] He was a perceptive sociological observer of the political systems and customs of the Amazonian and Andean tribes amongst whom he journeyed. Those familar with Spruce's *Notes of a botanist on the Amazon and Andes* will know that he was not only a distinguished botanist, but also a notable anthropologist,[34] linguist (French, Spanish and Portuguese), geologist and geographer.[35]

Never a robust man, Spruce's physical limitations were outweighed by his great strength of character. Few travellers have shown greater fortitude, endurance and unflagging dedication to their mission in the face of prolonged privations and hardships. The breadth of his interests,[36] the detailed accuracy of his observations[37] and the meticulous recording of all that he saw were phenomenal; he was also an accomplished artist and sketched native villages and the country through which he travelled, making maps of previously unexplored rivers. Nothing appears to have escaped his attention and capacity for orderly documentation.

Spruce was also talented in other ways.[38] A small hymnal entitled *The Welburn Appendix of Original Hymns and Tunes* by the Rev. James Gabb, Rector of Bulmer (another satellite village on the Castle Howard estate), with musical accompaniments edited by S.S. Wesley, cites the place and date of composition in the Preface as Welburn, Castle Howard: Easter 1875, and acknowledges, amongst others,

> *"Dr R. Spruce the distinguished naturalist and resident of this village for Tune No. 84 written before he entered on his travels in South America."*

This statement indicates that the tune, named 'Raywood' (after a wood adjacent to Castle Howard), was composed in or before 1849. Other than references to him as a musician by Stabler[39] and Wallace,[40] there are apparently no other contemporary references to Spruce's competence as a musician nor any allusion to what instrument or instruments he played, though a later anecdotal account tells of his playing the fiddle in old age.[41]

From 1864 until his death, Spruce gave help and advice to many bryologists throughout the world, mainly through the lively correspondence he maintained. His letters are notable both for their calligraphy and for their clarity of expression.[42] He was objective in his letters and writings and never sought to impress by dramatising his experiences; one's admiration for his achievements and prodigious industry, both during his travels and under the disabilities he suffered after his return, is enhanced by the absence of any hint of self-pity.[43] His close friend Matthew Slater wrote of him as

> *"courteous and dignified in manner but with a fund of quiet humour which rendered him a most delightful companion. He possessed in a marked degree the faculty of order, which manifested itself in the unvarying neatness of his dress, his beautifully regular handwriting and the orderly arrangement of all his surroundings. Whether in a native hut on the Rio Negro or in his little cottage in Yorkshire, his writing-material, his books, his microscope, his dried plants, his stores of food and clothing - all*

had their proper place, where his hand could be laid upon them in a moment. It was this habit of order, together with his passion for thoroughness, that made him so admirable a collector."

Spruce never married. He died at Coneysthorpe on 28 December 1893, aged 76, from the effects of influenza which his already weakened body could not overcome, and was buried beside his father and mother in the quiet churchyard at Terrington. Further biographical details are to be found in Stabler (1894), Wilkinson (1907), Wallace, in Spruce (1908), Sheppard (1909), Schultes (1953, 1968, 1983), Scott (1961), Sledge (1971), Seaward (1980) and Sledge and Schultes (1988).[44] Numerous flowering plants bear his name, two at generic level, *Sprucea* Benth. and *Sprucella* Stephani; of the three British bryophytes named after him, only *Marsupella sprucei* (Limpr.) H. Bern and *Orthotrichum sprucei* Mont. remain. For Richard Spruce there were few official honours - an honorary degree of Doctor of Philosophy from Dresden, an honorary fellowship of the Royal Geographical Society (1866), an honorary life membership of the Yorkshire Naturalists' Union (1891) and an Associateship of the Linnean Society (1893) - and two meagre pensions, £50 each per annum, from the British and Indian governments, awarded in 1865 and 1877 respectively. The four known portraits of him (two as line drawings) have been published on several occasions; examples of each appear in papers by Schultes.[45]

In 1970, Arthur Sledge and Richard Schultes made an international appeal in the botanical journal *Taxon* for funds to place a memorial plaque above the door of the cottage in Coneysthorpe where Spruce spent the last seventeen years of his life. The late George Howard of Castle Howard was well aware of Spruce's history and achievements and kindly agreed to the project. The appeal was supported by donations from twelve countries which more than covered the cost of the grey-green Westmoreland slate plaque with white lettering (Figure 3). On 3 September 1971, a ceremony was held at Coneysthorpe, presided over by George Howard, at which Richard Schultes unveiled the plaque. Representatives of the Yorkshire Naturalists' Union, the Yorkshire Naturalists' Trust, the Yorkshire Philosophical Society and the British Bryological Society were present at the ceremony. This plaque, unveiled more than a century after Spruce's travels in South America, appropriately commemorates his memory and his love of the peaceful tranquillity of rural Yorkshire to which he returned, after years in the Amazonian wilderness, to devote his remaining years to working on his bryological collections. Part of the fund was used to clean and re-letter Spruce's white marble scroll tombstone at Terrington: its simple inscription is an enduring memorial to an unassuming yet multi-talented man (Figure 4); the substantial sum remaining in the fund was donated, most appropriately, to the Margaret Mee Amazon Trust.[46]

In so short a compass it is hardly possible to do justice to the magnitude of the achievements and depth of scholarship of Richard Spruce. The wealth of information presented in the chapters which follow makes good this deficiency.

Fig. 4. Richard Spruce's tombstone, Terrington.

NOTES

(1)   Porter (Ch. 5) and Dickenson (Ch. 6) in *Richard Spruce (1870-1893), botanist and explorer* (eds. M.R.D. Seaward and S.M.D. FitzGerald), Kew, 1996 [hereafter 'this volume'], 51-63 and 65-80.

(2)   R. Spruce, *Notes of a botanist on the Amazon and Andes* (ed. A.R. Wallace), London, 1908, 2 vols. [Reprinted edition with a new foreword by R.E. Schultes, New York, 1970, 2 vols.]

(3)   For example, FitzGerald (Ch. 26), this volume, 299-302.

(4)   M.R.D. Seaward, 'Richard Schultes and the botanist-explorer Richard Spruce (1817-1893)', in *Festschrift for Richard Schultes*, Portland, OR: Dioscorides Press (in press).

(5)   Spruce, op. cit. (2), xlix.

(6)   Seaward (Ch. 27), this volume, 303-314.

(7)   FitzGerald (Ch. 26), this volume, 299-302.

(8)   Schultes (Ch. 1), Pearson (Ch. 2) and Spruce (Ch. 3), this volume, 15-25, 27-35 and 37-39.

(9)   R. Spruce, 'The Musci and Hepaticae of Teesdale', *Transactions of the Botanical Society of Edinburgh* (1846), **2**, 65-89.

(10)   M.R.D. Seaward, 'Two letters of bryological interest from Richard Spruce to David Moore', *Naturalist* (U.K.) (1980), **105**, 29-33.

(11)   R. Spruce, 'A list of the Musci and Hepaticae of Yorkshire', *Phytologist* (1845), **2**, 147-157.

(12)   R. Spruce, 'The Musci and Hepaticae of the Pyrenees', *Transactions of the Botanical Society of Edinburgh* (1850), **3**, 103-216.

(13)   G. Stabler, 'Obituary notice of Richard Spruce, PhD.', *Transactions and Proceedings of the Botanical Society of Edinburgh* (1894), **20**, 99-109.

(14)   Schultes (Ch. 1) and Prance (Ch. 8), this volume, 15-25 and 93-121

(15)   Field (Ch. 21), Edwards, (Ch. 22), Parnell (Ch. 23), Hackney (Ch. 24) and Lamond (Ch. 25) this volume, 245-264, 265-279, 281-286, 287-293 and 295-297.

(16)   Porter (Ch. 5) and Dickenson (Ch. 6), this volume, 51-63 and 65-80.

(17)   W.A. Sledge and R.E. Schultes, 'Richard Spruce: a multi-talented botanist', *Journal of Ethnobiology* (1988) **8**, 7-12; Naranjo (Ch. 13), this volume,163-170.

(18)   Drew (Ch. 12), this volume, 157-161.

(19)   For example, Vreeland (Ch. 17), this volume, 197-213.

(20)   Henderson (Ch.16), this volume, 187-196.

(21)   R. Spruce, 'Palmae Amazonicae, sive enumeratio palmarum in itinere suo per regiones Americae aequitoriales lectarum', *Journal of the Linnean Society, Botany* (1869), **11**, 65-183.

(22)   Stotler (Ch. 9), this volume, 123-140.

(23)   R. Spruce, 'Hepaticae Amazonicae et Andinae', *Transactions and Proceedings of the Botanical Society of Edinburgh* (1884-1885), **15**, 1-588.

(24)   Richards (Ch. 11), this volume, 151-155.

(25) Gradstein (Ch. 10), this volume, 141-150.

(26) G. Sayre, 'Cryptogamae exsiccatae - an annotated bibliography of exsiccatae of algae, lichenes, hepaticae, and musci. V. Unpublished exsiccatae. I. Collectors', *Memoirs of the New York Botanical Garden* (1975), **19**, 277-423.

(27) Ibid., 401-402.

(28) W. Mitten, 'Musci Austro-americani', *Journal of the Linnean Society, Botany* (1869), **12**, 1-659.

(29) Ewan (Ch. 4), this volume, 41-49.

(30) Dickenson (Ch. 6) and Field (Ch. 21), this volume, 65-80 and 245-264.

(31) Romero (Ch. 14), this volume, 171-182.

(32) Romero and Berry (Ch. 15), this volume, 183-185.

(33) Sledge and Schultes, op. cit. (17).

(34) Reichel-Dolmatoff (Ch. 20), this volume, 239-243.

(35) Smith (Ch. 18), this volume, 215-226.

(36) Schultes (Ch. 1), this volume, 15-25.

(37) Madriñan (Ch. 18), this volume, 215-226.

(38) Sledge and Schultes, op. cit. (17).

(39) Stabler, op. cit. (13).

(40) Spruce, op. cit. (2), xlii.

(41) Sledge and Schultes, op. cit. (17).

(42) Seaward,. op. cit. (10).

(43) W.A. Sledge, 'Richard Spruce', *Naturalist* (U.K.) (1971), **96**, 129-131.

(44) Stabler, op. cit. (13); H.J. Wilkinson, 'Historical account of the Herbarium of the Yorkshire Philosophical Society and the contributors thereto. Richard Spruce', *Annual Report Yorkshire Philosophical Society* (1907), 59-67; Spruce, op. cit. (2); T. Sheppard, 'A Yorkshire botanist. Richard Spruce (1817-93)', *Naturalist* (U.K.)(1909), **34**, 45-58; R.E. Schultes, 'Richard Spruce still lives', *Northern Gardener* (1953), **7**, 20-27, 55-61, 87-93, 121-125; R.E. Schultes, 'Some impacts of Spruce's Amazon exploration on modern phytochemical research', *Rhodora* (1968), **70**, 313-339; R.E. Schultes, 'Richard Spruce: an early ethnobotanist and explorer of the northwest Amazon and northern Andes', *Journal of Ethnobiology* (1983), **3**, 139-147; L.I. Scott, 'Bryology and bryologists in Yorkshire', *Naturalist* (U.K.) (1961), **86**, 155-160; Sledge, op. cit. (43); Seaward, op.cit. (10); Sledge and Schultes, op. cit. (17).

(45) R.E. Schultes, 'Richard Spruce still lives', *Northern Gardener* (1953), **7**, 21; R.E. Schultes, 'Some impacts of Spruce's Amazon exploration on modern phyto-chemical research', *Rhodora* (1968), **70**, 314; R.E. Schultes, 'The history of taxonomic studies in *Hevea*', *Regnum Vegetabile* (1970), **71**, 271.; R.E. Schultes, 'An unpublished letter by Richard Spruce on the theory of evolution', *Biological Journal of the Linnean Society* (1978), **10**, 160.

(46) Stiff (Ch. 7), this volume, 81-91.

Richard Spruce, the man

Richard Evans Schultes

Richard Spruce, the man

Richard Evans Schultes
Harvard Botanical Museum
Cambridge, Mass, USA

> *"To Richard Spruce the Amazon appeared as a monstrous tree. Its tribu-taries, its streams, its brooks were its boughs; the twigs, the branches of its ascending and spreading ramifications. His mind saw that dark region where the river-branches lost themselves as an impenetrable den-sity of green. One enters these green mansions and one vanishes, and vanish Spruce did on the edge of a primeval forest ..."*
>
> Victor W. von Hagen[1]

Many of us have a tendency quite properly to review all of the many and diverse outstanding accomplishments of Richard Spruce; but what kind of person was this extraordinary man?

Spruce deserves to be known both as a man and as a scientist, for his research and discoveries have benefited not only science but mankind in all cor-ners of the earth. Yet, though his achievements might be lauded in such widely separated countries as Brazil, Colombia, Ecuador, Peru, Venezuela, Ireland, Ceylon, Malaysia, Singapore, England, and France, his name – even in his native Yorkshire – goes unremembered beyond a limited circle of scientists or natural history enthusiasts.

The botanist who works in the New World tropics is frequently surprised that so few know of Richard Spruce and his accomplishments. Those of us who have travelled and collected where Spruce did – the northwest Amazon and the Ecuadorian and Peruvian Andes – find it even harder to believe that this obscuri-ty should exist.

Spruce could never have written a book about himself – he was too self-effacing and humble. However, he was an avid correspondent and kept extraordinarily detailed field notes. Fortunately, his famous colleague and admirer, Alfred Russel Wallace, took time from a busy life to edit Spruce's let-ters, field notebooks and various documents and assemble from them a truly great tribute to him: the two-volume *Notes of a botanist on the Amazon and Andes*, published in 1908, fifteen years after the death in 1893 of this intrepid plant explorer.[2] A work which has kept its appeal through the years, it repre-sents one of the most convincing accounts of a true self-directed search for knowledge. For many years now it has been almost unobtainable, even its fac-simile edition of 1970. In the main, Wallace let Spruce speak for himself in the compilation, doubting that anyone could excel Spruce's depth of perception, Yorkshire wit, forthright style in recounting his experiences and expounding his philosophy. It has truthfully been said:

"... everything is to be found in Spruce, and the temptation to quote him is irresistible".[3]

In Spruce, we find perhaps the greatest contrasts ever known in a plant explorer. A man of extremely delicate health and plagued by chronic ailments, he betook himself to one of the world's wildest and least known areas of jungle and high mountains, to spend fifteen years in physical work, constantly exposed to tropical climates and diseases, often with insufficient diet and lacking even rudimentary comforts. A classical scholar with a cultured and scientific mind, he divorced himself from all centres of accustomed civilisation and lived for long periods amongst Indians and unlettered settlers. A superb correspondent, he plunged into regions where, for months on end, he received not a letter nor a newspaper. A scientist trained to handle masses of minute detail, he could cope with the endless gross problems attendant on the organisation and execution of cumbersome trips by canoe and horseback for months' or even years' duration. A researcher whose first love concerned diminutive land plants – mosses and liverworts – he collected, studied, named and described some of the loftiest tropical trees and lianas, and discovered hundreds hitherto unknown to science. A poor man throughout his life, he laid the foundations of our knowledge of *Hevea* rubber trees and contributed, through his work on the quinine trees, to the creation of great plantations in the British colonies. His labour helped to make the price of this medicine lower and available to millions of the poor in tropical malarial countries, like India. A mild-mannered and dignified person, he accepted the dangers that he met and, more than once, had to take extreme measures to defend his very life.

Perhaps most astonishing of all – this naturalist who abhorred the philosophy that nothing not of use to man was worthy of study, nevertheless filled his letters and field notebooks with ethnobotanical notes and personal studies on all manner of economic plants: gums and resins, fibres, foods, drugs, narcotics and stimulants, oils, dyes and timbers. He lived and worked with meticulous care, evident not only in the neatness of his dress and handwriting, but in the way that he always, even when desperately ill, insisted on doing all the labour connected with the collecting and preservation of his specimens, never relying completely on hired help. A serious man, he nevertheless had a gentle sense of humour, even when nothing proceeded as planned. A botanist at heart, he was interested in all that he encountered and made notes and drawings of everything: Indian faces and houses, panoramas, plants, rapids, strange rock formations, petroglyphs and many other curiosities. Above all, he possessed a scientific mind, as evidenced by his simple statement in a letter written in 1851 and preserved by the Yorkshire Philosophical Society:

"Then there is the greatest of all pleasures to the naturalist, however some utilitarians may affect to undervalue it, that of discovering new species, of dotting in (as it were) new islands on the map of nature, and, in some cases, of even peopling continents that appear to be deserts".[4]

In politics and religion, Spruce was liberal, always expressing his sympathy for the downtrodden. One aspect of his religious feelings can be gleaned from his words to Daniel Hanbury in 1873, concerning his beloved hepatics:

"I like to look on plants as sentient beings, which live and enjoy their lives – which beautify the earth during life and after death may adorn my herbarium. When they are beaten to pulp or powder in the apothe-cary's mortar, they lose most of their interest for me. It is true that the Hepaticae have hardly as yet yielded any substance to man capable of stupefying him or of forcing his stomach to empty its contents, nor are they good for food; but if man cannot torture them to his uses or abuses, they are infinitely useful where God has placed them, as I hope to live to show, and they are, at the least, useful to, and beautiful in, themselves – surely the primary motive for every individual existence".[5]

It was his boyhood rambles in his native countryside that permitted him to compile at age 16 a list of 403 species of plants of Ganthorpe and several years later a list of the *Flora of the Malton District* with 485 species. It was at this time that he acquired his life-long love of bryophytes –

"the joy of his early manhood and the consolation of his declining years".[6]

His first venture outside England was a collecting trip to the Pyrenees in 1844 which he planned to finance by selling sets of herbarium specimens of his plants. A famous French guide book had stated categorically that mosses did not exist in the Pyrenees, although a French bryologist had published a list of 169 species; Spruce's collection yielded 498, of which 17 were new to science and 73 new to the Pyrenees. He later published a scholarly 114-page work entitled *The Musci and Hepaticae of the Pyrenees* (1850).[7]

Returning to England, he continued collecting and selling specimens to foreign herbaria. His preoccupation about the future, however, is evident even in his customary calmness, for in 1846 he wrote in a letter to William Borrer:

"I yearn to be independent, and I hope the next time I go out it will be to settle in some comfortable office, and in the meantime what my hand has found to do I will do with all my heart, for my heart is in it".[8]

His next "comfortable office" was the Amazon jungle! Typically for Spruce, he took his new place of work immediately to heart. He wrote:

"The largest river in the world runs through the largest forest. Fancy ... two millions of square miles of forest, uninterrupted save by the streams that traverse it ... The natives ... think no more of destroying the noblest trees ... than we the vilest weed; a single tree cut down makes no greater gap and is no more missed than when one pulls up a stalk of groundsel or a poppy in an English cornfield ... I realised my idea of a primeval forest. There were enormous trees crowned with magnificent foliage, decked with

fantastic parasites ... Violets ... grow to be trees; milkworts are represented by woody climbers climbing to the tops of the loftiest trees ...". [9]

Spruce fell in love with the forest and soon became thoroughly familiar with it – so that in 1858, he wrote in a letter an estimate so modern in outlook that even today, some botanists are loath to accept it:

"I have lately been calculating the number of species that yet remain to be discovered in the great Amazonian forest ... there still should remain some 50,000 or even 80,000 species undiscovered. [10]

How exceedingly foresighted this estimate!

His years on the Rio Negro of Brazil were plagued with constant adversity. A lesser man would not have kept his cheerful disposition so long. This region is still today the home of hunger. Often he had to forego plant collecting to take his shotgun to get food, several times procuring only a parrot. An inveterate letter writer, his greatest joy – next to the discovery of a new plant - came from the all-too-infrequent post. Sometimes in his letters, he seemed almost in despair, as when he wrote:

"My last dates from England are a year old. Neither newspapers nor anything else ever reaches me now. I seem to have taken my last leave of civilisation". [11]

In one of the houses in São Gabriel which he used as a base:

"the thatch is stocked with rats, vampires, scorpions, cockroaches and other pests to society; the floor (being simply mother earth) is undermined by [leaf-cutting] ants, with whom I have had some terrible contests. In one night, they carried off as much farinha as I could eat in a month; then they found my dried plants and began to cut them up and carry them off ... Then the termites ... have covered ways along every post and beam. But the greatest nuisance at São Gabriel is one I had not foreseen ... When a man commits a crime ... he is enlisted [as a soldier] and marched off to ... frontier posts ... of the fourteen [here] at least half of them are murderers. Judge with what security I can leave my house ... [it] has already been twice entered ... and about two gallons of spirits, a quantity of molasses and vinegar and some other things have been stolen ...". [12]

Ill and following a series of tremendous disappointments, our mild mannered naturalist still kept his equanimity; at the highest point on the Uaupés River in Brazil near the Colombian border, at the time of year when the trees were coming into full flower, he wrote back to Kew:

"... as I scanned the lofty trees with wistful and disappointed eyes ... until I could no longer bear the sight, and covering up my face with my hands,

I resigned myself to the sorrowful reflections that I must leave all these fine things ... But I find that it will be necessary to cross páramos rugged and inhospitable ... and ... risk myself among wild and fierce Indians, so I fear its exploration must be left to someone younger and more vigorous than myself". (13)

When he returned to Manáus after four years on the Rio Negro and the adjacent Orinoquia of Venezuela, the rubber boom had gripped the Amazon. In his criticism of the enslavement, torture and murder of Indian rubber tappers and the annihilation of whole cultures and tribes by "rubber barons", his innate love of his fellow man showed: Spruce observed that, although the oppressive government *fazendas* [forced labour farms] and the seizure of Indians had been outlawed,

"yet the practice still exists and is carried out. I speak of this with certainty, because since I came up the Rio Negro two such expeditions have been sent ... to make pegas [seizures] among the Karapaná Indians". (14)

The many descriptions and analyses of the early years of this nefarious business are basic to any modern economic, humanitarian, sociological and historical study of present conditions, for, in some parts of the Amazon, mistreatment of natives is still rife.

Yearning to see the forests in the eastern Amazon slopes of the Andes, he boarded a steamer and, as it travelled upstream, he was able to enjoy a much-needed rest, arriving finally in Iquitos, Peru on 1 April 1855. It was this leisurely trip that Spruce needed; although, aboard a craft that he did not command, he was impatient to stop and collect the increasingly interesting flora, much of which was new to him!

Eventually, on 21 June 1855, he reached the small town of Tarapoto, where he spent two years amassing a vast collection, including 250 species of ferns, and where he rested, with an enjoyable climate, good food and friendly towns-people. To use the word *resting*, whilst making such quantities of collections seems really to be paradoxical, but his stay in Tarapoto did rejuvenate him, partly because of the extensive collections of bryophytes. He wrote:

"My chagrin at the delay was somewhat lessened by the circumstance of finding myself in the most mossy place I had yet seen anywhere ... and I found reason to forget ... my troubles in the contemplation of a simple moss". (15)

The delays also enabled him to make important observations and notes on curious aspects of the forest trees, many of which were new to him. It was his extraordinary collections in Tarapoto that undoubtedly engendered the idea of eventually writing his later masterly publication on the hepatics in which he

"almost single-handedly established neotropical hepaticology on a sound basis". (16)

And, despite all the inconveniences and sufferings, his determination to continue his journey never wavered.

During this stay in Tarapoto he changed his plans in order to continue his work in Ecuador instead of Peru which was in revolution. It was also at this time, in 1857, that he was commissioned by Her Majesty's Secretary of State for India to proceed to the *Cinchona* forests of the Andes to study the strains richest in the quinine alkaloid and to collect seed and living material for the establishment of plantations of this medicinal tree in British colonies in Asia. What a mighty commission it was – to lay the scientific basis of an ambitious new plantation industry! Proof that, thanks to his achievements of ten years in the remote verdure of South America's rain-forests, this modest self-trained botanist from Yorkshire was now recognised as a leading figure in world botany.

His journey, by canoe and on foot from Tarapoto up to Baños at the base of Tunguragua Volcano turned out to be one of the most demanding trips he had ever undertaken, but he faced it with his accustomed stoicism and with happy anticipation of the marvels of nature that lay ahead. It was a difficult journey of 100 days, 90 of them in perilous rapids; the greatest peril was the swelling rivers with whole trees floating downstream. Of one night, he wrote:

> *"I had slight hope of living to see the day, and I shall ever feel grateful to those Indians who, without any orders ..., stood through all the rain and storm of that fearful night, relaxing not a moment ...".*[17]

To get across rivers, he once had to construct a makeshift bamboo bridge so fragile that he had to reduce weight by disposing of baggage, particularly his supply of paper needed for plant collecting.

> *"Those who have escaped death by hunger or drowning may understand what a load was taken off my heart when we had all got safely across ..., although I had been obliged to abandon so many things which were to me more valuable than money".*[18]

He found Baños to be *"a poor little place of about a thousand souls"* with plentiful food (including bread for the first time in many days) and congenial inhabitants; he was rejuvenated in body and renewed in spirit from his stay there.

From Baños, he finally arrived in Ambato at the base of the great Mount Chimborazo; here he remained and collected extensively until early 1860, when he suffered a nearly complete physical breakdown. In spite of *"unspeakable ... physical pain"*, however, he began his epoch-making work on the quinine trees. It took all his friendly persuasiveness to get permission from the Ecuadorian authorities to engage in this work on the *Cinchona* collecting, but by August of that year, although he was nearly a physical wreck, he was able to carry on his studies and collections in the richest area of the finest types of *Cinchona*. He listened to the native bark collectors but verified all folk-information concerning the trees, and slowly he amassed more data than anyone had ever gathered on this life-saving tree. Doing all of the work entirely by himself, he eventually collected 2500 fruits with approximately 100,000 seeds which he shipped from the Pacific port of Guayaquil and which arrived safely in India. He then returned to the highlands to prepare for shipment his 637 seedlings which he had planted; these went down to Guayaquil by raft and were shipped at the beginning of 1861.

During these months of work in the high Andes of Ecuador, Spruce's health had seriously deteriorated. Any ordinary individual, worn out and ill, would have returned at once to England after eleven years in South America's tropical rain forests and high Andes; his Ecuadorian friends and British colleagues urged him to return to an easier life of retirement, but entranced with the dense forests on Ecuador's Pacific slopes, he remained until 1864, collecting the general flora of species wholly new to him.

Having travelled and worked for fifteen years in the wildest of regions – some so inaccessible that only now, after more than 140 years, are they again being penetrated by botanists – he returned to England, and spent his declining years in a cottage in the tiny hamlet of Coneysthorpe on the Castle Howard estate near his birthplace. Here he lived in the simplicity always characteristic of him, wherever he happened to be. He had to live frugally on a modest pension of £100 a year from the governments of Great Britain and India, given in recognition of his achievements in procuring *Cinchona* plants for the vital new quinine industry. Shortly before leaving the Andes to return to England, he lost all his savings, a paltry £700, when the Ecuadorian institution where he had deposited the funds went bankrupt. This shock was almost too much for him – already completely broken in health – to bear; but he bounded back, even from this unexpected calamity.

> "He ... received that pittance of a pension .... He had really done it the hard way, and he then died the hard way, too, after years and years of debilitating illness. The Cinchona plantations and the rubber trees were growing well when he, so much the pioneer in all this enterprise, gave up the struggle in 1893". [19]

Such a man was Richard Spruce, whose greatness as a scientist has been enhanced not only by his achievements but by his humbleness.

His residence in Coneysthorpe marked the end of his travels, as several years after his return he gradually became so ill that even short walks were barely possible. Sitting at his microscope for any length of time brought on intestinal bleeding which, after numerous mis-diagnoses, the physicians decided that it was due to "stricture of the rectum". He wrote to a friend that the physicians:

> " ... found it so much easier to hide their ignorance under accusations of hypochondria and to prescribe brandy and water every three hours". [20]

He never lost his humour, however, for he wrote to another friend:

> "One day last week, a dentist relieved me of four teeth, and I belong now to the genus Gymnostomum [a bryophyte genus], but by the time you come over I hope to have developed a complete double peristome" [tooth-like appendages at the opening of a moss capsule]. [21]

During his years of retirement, he kept up a lively correspondence, much of it with bryologists, and, while still undoubtedly in spirit wandering his old trails in the Amazon and Andes and greeting the mosses, hepatics and trees, he produced thirty-four scientific papers and several books, reviewing 700 species and varieties, 500 of which he had collected and 400 of which he had described.[22]

It is difficult, if not impossible, to decide which of Spruce's great contributions to ethnobotany has been the most significant. Certainly one of the most perceptive accomplishments must be his identification of the widely employed sacred Indian hallucinogenic plant in western Amazonia. There were numerous historical references to *ayahuasca*, *caapi* and *vajé* in the writings of early travellers and missionaries, but its botanical source was not known until 1853 when Spruce attended a native ceremony amongst the Tukanos in the Rio Uaupés and later sought out the plant, a liana cultivated in an Indian garden. He collected the type specimen from this plant and identified it as a new species of the Malpighiaceae, naming it *Banisteria caapi*, now, as a result of nomenclatural changes, known as *Banisteriopsis caapi*.

Spruce's inquisitive thoroughness is evident in his introduction to this drug. His own description of his experience in the Indian dance is best told in his own inimitably amusing words. He tried one draught of *caapi*, finding it bitter and choking; immediately:

> *"the ruler of the feast, desirous ... that I shall taste all his delicacies at once, came up with a woman bearing a large calabash of manioc beer, of which I must ... take a copious draught, and as I knew the mode of its preparation, it was gulped down with secret loathing. Scarcely had I accomplished this feat when a large cigar two feet long and as thick as the wrist was put lighted into my hands, and etiquette demanded that I should take a few whiffs of it – I who had never ... smoked a cigar or pipe of tobacco. Above all this, I must drink a large cup of palm wine ..."*.[23]

The extraordinary aspect of Spruce's discovery and identification of *caapi* is that, recognising that the malpighiaceous family was not known to have bioactive chemical constituents, he collected for chemical analysis a number of pieces of the trunk of the great liana from which his flowering herbarium specimens came. This foresight was hardly known amongst plant collectors of that period. There was usually little liaison between botanical explorers and laboratory chemists![24] He wrote:

> *"I obtained a good many pieces of stem ... The man who took the box ... was seized for debt when about half way down the Río Negro ... My boxes were thrown aside in a hut with only the damp earth for floor and remained there many months ..."*;

eventually a friend found them and they were sent to Kew Gardens. When they arrived in England, they were somewhat injured by dampness and mould and, as Spruce himself felt:

> *"... the caapi would presumably have quite lost its virtue ... and I do not
> know that it was ever analysed ...".*[25]

In 1968, these pieces of stem were located at Kew and finally analysed, with
the result that alkaloids were found to be 0.4%; a recently collected specimen
was found to contain 0.5% alkaloids.[26]

Spruce was multi-talented. Not only was he a superb botanist and philosophical thinker on evolution in the Plant Kingdom[27] but an anthropologist,
geologist, geographer, linguist and an accomplished artist who, during his field
work in the Amazon, made beautiful pencil sketches of plants, panoramas, petroglyphs, and Indians.

He had another talent not frequently realised: in Spruce's years of "retirement" in Coneysthorpe, he played his "fiddle" every evening. Even before his
departure for the Amazon in 1849, he had composed a hymn which was published in an Anglican hymn-book;[28] in the Amazon, he enjoyed notating the
simple monophonic melodies played by the pan-pipes of the Indian dances and
ceremonies. He typified the all-round scientist and gentleman of culture that
unfortunately today is so sorely missed in over-specialisation and compartmen-
talisation of the sciences.

Richard Spruce still fires the heart and shapes the thoughts of many a plant
explorer, carrying forward his great unfinished work in the Amazon.[29]

NOTES

(1)   V.W. von Hagen, *South America called them: explorations of the great naturalists,
Charles-Marie de la Condamine, Alexander von Humboldt, Charles Darwin, Richard
Spruce,* 1949, 291-374.

(2)   R. Spruce, *Notes of a botanist on the Andes and Amazon* (ed. A.R. Wallace),
London, 1908, 2 vols. [Reprinted edition with a new foreword by R.E. Schultes,
New York, 1970, 2 vols.]

(3)   C. Sandeman, 'Richard Spruce. Portrait of a great Englishman', *Journal of the
Royal Horticultural Society* (1949), **74**, 531-544.

(4)   R.E. Schultes, 'Plantae Austro-Americanae VII. De festo seculari Ricardi Sprucei
America Australi adventu commemoratio atque de plantis principaliter Vallis
Amazonicis diversae observationes', *Botanical Museum Leaflets (Harvard
University)* (1951), **15**, 29-78.

(5)   Spruce, op. cit. (2), vol.1, xxxix.

(6)   R.M. Schuster, 'Richard Spruce (1817-1893): a biographical sketch and appreciation', *Nova Hedwigia* (1982), **36**, 199-208.

(7)   R. Spruce, 'The Musci and Hepaticae of the Pyrenees', *Transactions of the
Botanical Society of Edinburgh* (1850), **3**, 103-216.

(8)   Spruce, op. cit. (2), vol.1, xxxii.

(9)   Ibid., 256, 3-43.

(10) Ibid., vol.2, 208.

(11) Ibid., vol.1, 299.

(12) Ibid., 293.

(13) Ibid., vol. 2, 208.

(14) J. Hemming, *Amazon frontier: the defeat of the Brazilian Indians*, London, 1987, 146-147, 273-274; Spruce, op. cit. (2), vol. 1, 355.

(15) Spruce, op. cit. (2).

(16) Schuster, op. cit. (6).

(17) Spruce, op. cit. (2), vol. 2, 122.

(18) Ibid., 157.

(19) A. Smith, 'Explorers of the Amazon', *Viking* (1990), 269, 283, 291.

(20) Spruce, op. cit. (2), vol.1, xxxiv.

(21) Ibid., xxxvi.

(22) Schuster, op. cit. (6).

(23) Spruce, op. cit. (2), vol. 2, 420.

(24) R.E. Schultes, 'Some impacts of Spruce's Amazon explorations on modern phytochemical research', *Rhodora* (1968), **70**, 313-339.

(25) Spruce, op. cit. (2), vol. 2, 420.

(26) R.E. Schultes, B. Holmstedt and J.-E. Lindgren, 'De plantis toxicariis e Mundo Novo tropicale commentationes III. Phytochemical examination of Spruce's original collection of *Banisteriopsis caapi*', *Botanical Museum Leaflet (Harvard University)* (1969), **22**, 121-132.

(27) R.E. Schultes, 'An unpublished letter by Richard Spruce on the theory of evolution', *Biological Journal of the Linnean Society* (1978), **10**, 159-161; R.E. Schultes, 'Still another unpublished letter from Richard Spruce on evolution', *Rhodora* (1987), **89,** 101-106.

(28) W.A. Sledge and R.E. Schultes, 'Richard Spruce: a multi-talented botanist', *Journal of Ethnobiology* (1988), **8**, 7-12.

(29) R.E. Schultes, 'Richard Spruce still lives', *Northern Gardener* (1953), **7**, 20-27, 55-61, 87-93, 121-125; reprinted, *Hortulus Aliquando* (1978), **3,** 13-47.

FURTHER REFERENCES CONSULTED

W.A. Sledge, 'Richard Spruce', *Naturalist* (U.K.) (1971), **96**, 129-131.

V.W. von Hagen, 'The great mother forest: a record of Richard Spruce's days along the Amazon', *Journal of the New York Botanical Garden* (1944), **45**, 73-80.

2

Richard Spruce, the development of a naturalist

M.B. Pearson

Richard Spruce: the development of a naturalist

M.B. Pearson

Hunters Croft, Haxey, Doncaster, UK

Introduction

Henry Walter Bates, Alfred Russel Wallace and Richard Spruce are the three great Victorian naturalists who made outstanding contributions to our understanding of Amazonian natural history. Bates and Wallace left England in their early twenties and were to spend eleven and four years respectively collecting in the region. When Spruce arrived at Belém in 1849, at the age of 31, he was already a seasoned collector and was to spend a total of fifteen years travelling in the Amazon and Andes. Others have rightly focused on Spruce's achievements in South America. In comparison little has been written about his early life, his family background and his development as a naturalist.

Family background

From an examination of parish and census records, and details provided by Spruce's extensive correspondence and Wallace's biographical sketch[1] it is possible to piece together an outline of the family background.

Spruce's father, also named Richard, was the teacher in the village school of Ganthorpe. He had been born in Strensall, a village on the outskirts of York. His grandfather, another Richard, was a farmer whose parents had come from the village of Weaverthorpe. Little is recorded about Ann Etty, Spruce's mother. She is said to have been a relative of William Etty, the well known painter, and to have possessed an *"exceedingly nervous temperament"*.[2]

The parish registers of Terrington, in North Yorkshire, show that Richard Spruce (the botanist) was born in Ganthorpe and baptised on 10 September 1817 by the curate, Robert Frere. Thus Richard was the first child; his sister, Anna, was born two years later. A second sister, Mary Ann, arrived in 1822 but survived for just two weeks. Then in early January 1828 Anna died at the age of 8, and the following year his mother died at the age of 38. The records provide no clues as to the cause of these deaths.

In 1832, less than 3 years later, Spruce's father remarried. The bride, Mary Prest, from the same parish, was 24 years old, the groom was 50; and the register shows that she was unable to sign her own name. This remarriage seems to have been resented by the younger Richard who, in a letter written twelve years later, expressed his outrage that his father had married a woman half his age.[3] Relations between Spruce and his stepmother were described as being *"no more cordial than usually exists between stepmother and stepson"*.[4]

In 1833, Hannah, the first child of this remarriage, was born, followed at regular intervals by seven more girls, their baptisms all recorded in the Terrington and Bulmer parish registers.

Early life

Spruce appears to have shown an early interest in botany and by the age of 16 he had drawn up a list of 403 species found around his home village. In 1837, just three years later, he had produced a manuscript (now lost), entitled 'List of the Flora of the Malton District' which contained 485 species of flowering plants. This expertise was soon acknowledged and several of Spruce's localities for the rarer plants were referred to in Henry Baines' *Flora of Yorkshire* published in 1840.

When Spruce was not roaming the countryside in search of plants he was studying and later assisting at his father's school. He was clearly a gifted child who demonstrated a mastery of the academic subjects of that time. The Latin that he learned proved invaluable later in producing the formal descriptions of the new species discovered in South America. In later correspondence he revealed that his two favourite books were John Bunyan's *The Pilgrim's Progress* and Daniel Defoe's *Robinson Crusoe*.[5] At about the age of 20, he left home to become a tutor at a school in Haxby, near York. This was followed, at the end of 1839, by a post as mathematics teacher at the Collegiate School in York. On leaving Ganthorpe, Spruce's horizons widened and he had the opportunity to travel further afield to botanise. From June 1841 he started to list his excursions in small notebooks which he continued to use throughout his later explorations of the Amazon and Andes. Most of the entries are simply details of place-names with occasionally a species of note recorded.

One of the earliest entries for the end of June 1841 (in the Linnean Society Library) entitled 'List of botanical excursions' records a 3-day excursion to the North York Moors, in the company of Henry Ibbotson, a fellow teacher and botanist from Ganthorpe. Later he wrote an account of it which was published in *The Phytologist* - his first published paper.[6] From the published account it is possible to retrace their steps in some detail and study the lengthy list of plants and their locations. By the end of their first day they had reached Whitstonecliffe and spent the night at the Hambleton Hotel. Next day they tramped on to Scawton, Rievaulx Abbey, Helmsley, Kirby Moorside and finally reached Pickering at 10 p.m. The last leg of the journey took them on to the Hole of Horcum and back to Ganthorpe by way of Newsham Bridge.

For the rest of the summer and autumn of that year the notebook shows that Spruce was mainly restricted to the area around York. However, in December he ventured further afield and spent a couple of days collecting in Wharfedale in the Yorkshire Dales. Again a brief note of the results was published in *The Phytologist*.[7] From this account it is clear that by this time Spruce was concentrating on bryophytes. Along a three mile stretch of the River Wharfe, from Bolton Abbey to Barden Tower, Spruce recorded a total of nineteen species of mosses, sixteen liverworts and eight lichens. The identification of species often depends on the microscopic examination of the specimens and Spruce relied on the bryological expertise of Thomas Taylor (1786-1848) in Co. Kerry in Ireland to clear up any doubts. In 1842 he spent two months during the summer with Taylor in Ireland, botanising and studying his extensive herbarium.

In the same year he also worked on his 'List of the Musci and Hepaticae of Yorkshire' (297 mosses and 76 liverworts) in which he recorded no less than 48 mosses new to the English flora and 33 others new to Yorkshire (published 1845).

2. Richard Spruce, the development of a naturalist

His next summer was devoted to an exploration of Teesdale, an area with a particularly interesting flora. Ever since the 17th century, when the naturalist John Ray had found the shrubby cinquefoil there, Upper Teesdale has been renowned for its flora. Despite the interest shown by botanists in this area, little attention had been paid to the bryophytes. Hooker's *British Flora* listed only a single species of moss from there and Baines' *Yorkshire Flora* contained only four Teesdale mosses. In just three weeks Spruce increased the records to 167 mosses and 41 liverworts and discovered seven species new to Britain - clearly a very successful holiday.

During this time Spruce continued to teach at the Collegiate School but devoted his free Wednesday and Saturday afternoons as well as his school holidays to the pursuit of botany. In a letter to William Borrer he wrote:

> *"As my work is to study the plants I collect & not merely to amass an extensive collection, my small amount of leisure obliges me to confine my Botanical pursuits within very narrow limits."*(8)

From reading his correspondence of this time one might conclude that Spruce led a very restricted social life. His letters contain little personal information beyond the botanical discoveries and discussions about identification. However, it is clear from his published papers that much of his botanising was done in the company of Henry Ibbotson and John Teesdale. Correspondence with the latter was used by Wallace in his editing of Spruce's *Notes of a botanist*,(9) but seems to have disappeared since. In a letter written twenty years later to Daniel Hanbury, another friend is mentioned: Elizabeth Teale Sigston, only child of the surgeon at Welburn and fifteen years old when Spruce was in York. According to Spruce her friends warned her against marrying *"a young man of some ability, but who was throwing himself away on so futile a pursuit as botany"*.(10) In fact Spruce was forced by other circumstances to make a choice between his profession of teaching and his love of botany. In a letter to the Sussex botanist William Borrer he broke the news:

> *".. intelligence came to me that the York Collegiate School was sold, & the establishment completely broken up. It wd. occupy a great deal of space to enter into a complete account of the transaction; but it may suffice to state that the Coll. School was started a few years ago in opposition to an ancient grammar school (St. Peter's) connected with the minster, & which though a very wealthy establishment (having an income of £1700 a year independent of pupils) had become very inefficient, from mismanagement or some other cause. The Dean & Chapter had been for the last 2 or 3 years devising ways & means for rendering it more prosperous, but without success, until they suddenly hit on the expedient of purchasing the premises of the Collegiate School, & transferring their establishment thither, a most effectual way of annihilating their rival! There are however several circumstances which are likely to lead to litigation, one of which is that the Directors of the Coll. School have sold the property without the consent of the remaining proprietors. However, this is of little moment to me, who am suddenly & without the least previous intimation thrown on my own resources. Another circum-*

stance against me is that my health still continues very precarious. I am at present under medical treatment & had two blisters applied to my chest 2 or 3 days ago, and I am ordered to apply more. In consequence of this I think I shall not attempt to engage any other situation, at least of a similar kind, for some months yet. I have long wished I could obtain some office of a less confining nature & which would leave me more leisure for cultivating Botanical pursuits, but I never saw how this could be done with my limited means. Any advice, my dear Sir, from you on this subject will be extremely acceptable."[11]

Further letters passed between Spruce and Borrer over the next few months. He applied for the Curatorship of the Birmingham Philosophical Institution but thought the duties too onerous for a salary of £80 a year. The Curator was expected to take care of the Museum, lecture to the members and give a course of at least forty lectures to a junior class throughout the year. This job compared unfavourably with the Yorkshire Philosophical Society which paid its Curator £200 per year to care solely for the Museum. It was clear that Spruce did not make a very pressing application and had neither hope nor expectation of being appointed.[12]

Fortunately Borrer came to his aid and contacted Sir William Hooker, the newly appointed Director of Kew. A plant agency in London and then the curatorship of some colonial botanical garden were successively discussed and in the end rejected. Finally, in December 1844, Hooker suggested that Spruce should travel to Spain as a plant collector and lent him George Bentham's account of his visit to the Pyrenees. This seems to have fired his enthusiasm for the idea; but then he began to have doubts as to the financial viability of the scheme and heard that a friend had recommended him for a teaching post at a school near Halifax with a salary of £100 a year.[13] However after further correspondence with fellow botanists, he was won over to the career of plant collector.

Plant collector

Preparations began in earnest; Spruce returned to York for lessons in conversational French, as the plan was for him to head for the Pyrenees and then, if safe, to proceed into Spain. He was to collect all botanical material of interest and then sell sets of the dried plants to museums, botanical gardens and interested individuals for their herbaria.

Just as all the arrangements were near completion disaster struck. In March 1845 there was an outbreak of scarlatina in Welburn where Spruce's father was now teaching and where the family lived. Four of Spruce's sisters caught the infection and he returned to the village to care for them. The burial register records the sad outcome. Sarah Jane, aged seven years, was the first to die. Diana, aged two years, was buried a week later on 30th March. Three days later Mary Ann was also buried, not having reached her fifth birthday. A fourth sister, either Hannah or Elizabeth, recovered.

Spruce witnessed the loss of three of his young sisters in quick succession, and in addition his father had a severe attack of erysipelas, and Spruce had to take over his duties at the local school until he recovered. After all this it must have been a welcome relief to Spruce to leave Yorkshire.

In the 19th century, the journey to the Pyrenees was no small undertaking. Travelling via London at the end of April 1845, Spruce took ship at Shoreham in Sussex and suffered a stormy passage to Dieppe which left him feeling very sea sick. He then travelled on to Paris where he spent four days before undertaking the forty-four hour coach ride south to Bordeaux. Then it was another lengthy ride on to Pau and the start of his botanical explorations of the Pyrenees. Here he met a Dr. Southby (c. 1800-1883), a wealthy Wiltshire landowner, who lived in France for the sake of his wife's health. Southby was to be Spruce's companion on many of his travels in the area.

Spruce's account of this year of travelling and collecting was read to the Botanical Society of Edinburgh in 1849:

> "I arrived at Pau ... ancient capital of Bearn, in the early part of May 1845, & my first herborization in the Pyrenees was made on 13th of the same month. My excursions comprised, besides the woods etc adjoining the town of Pau, the villages of Jurançon, Gelos, Rontignon, & Narcastet, lying on the southern bank of the Gave de Pau, with the valleys running up from them to the southward, among what may be called the skirts of the Pyrenees; & the village of Bilhères, lying south of the same river. From the 29th to the 31st were devoted to a visit to Oloron, at the entrance of the Vallée d'Aspe, along which runs one of the most frequented roads into Spain. On the 11th of June I again left Pau for St Sever, in the Landes, on a visit to Dr Léon Dufour, the eminent naturalist, where eight days were usefully spent in exploring the neighbouring Landes, especially those of Mugriet (Commune of Souprosse) a few miles distant from St Sever, & on the opposite side of the Adour. Returning thence to Pau, I again started on the 25th for Laruns, a little town lying about 20 miles southward. Here commenced my acquaintance with the real Pyrenees. My excursions included the Pic de Ger & the Montagne Verte; the Gorge de Hourat, to the Eaux Chaude, & watered by the Gave de Gabas; the Gave de Valentin ... & the Lameau of Bages.
>
> On 2nd August, accompanied by Dr Southby I crossed the central chain by the Fort de Cauterets to the baths of Penticosa in Aragon ... During my stay here of 5 wks I explored the whole of the magnificent Vallée du Lys (lateral to the valley of Luchou) with its four lakes and twenty four cascades.
>
> The autumn being unusually prolonged & the summits still clear of snow I undertook another expedition to the Basses Pyrenées & on 1st November proceeded again to Laruns where I remained until fairly driven away by the coming of winter ... Driven from the mountains my next destination was by way of Pau to Dax in the Landes where I arrived on 18th November. In the midst of almost unceasing rain I visited in this rich district the ophitic rocks of St Pandelon on the banks of the Luy (a tributary of the Adour), the chalk rocks of Tercis, & the woods of Sanbagnac & La Torte. Having devoted a fortnight to a re-examination of the neighbourhood of Pau I returned early in December to Bagnères to winter. In the Pyrenees as throughout nearly all the rest of Europe, the

winter of 1845-6 was remarkably mild & by the month of February the lower mountains were quite clear of snow ... in company with M. Philippe, an excursion of four days into the heart of the mountains, for the purpose of examining the back of the Pic de Mont-Aigu & the Vallée de Castelloubon. Even at that season we were able to reach an altitude of 7000 ft & might easily have gone higher but the ground at that height was frozen to the depth of several inches & the waterfalls were changed into sheets of ice.

Finally quitting Bagnères early in March, a last visit to Pau rendered my collection of the mosses of the Western Pyrenees still more complete; & in proceeding thence to Paris, two days spent at St Sever with the excellent Dufour afforded me rarities unobserved the preceding year."[14]

On his return to England in April 1846, Spruce took a holiday with William Borrer in Sussex and then got down to organising the sale of his Pyreneean collection. Herbarium specimens were distributed at a price of 30 shillings per hundred plants. At the same time Spruce organised his collection of mosses and liverworts and prepared a lengthy paper for publication. This work was interrupted on several occasions by the serious and protracted illness of his father. During 1847 and the following year Spruce again had to take over his ailing father's duties and run the village school.

Initially botanists such as George Bentham and George Walker-Arnott had advised that he would find few new bryophytes in the Pyrenees. The French authority Léon Dufour had listed a total of 156 mosses and 13 liverworts for the area. It comes as no surprise to find that by the time Spruce had finished he had raised the total number of species to 386 mosses and 92 liverworts. Eighteen months after his return he had completed his paper, which occupied 114 pages of the *Transactions of the Botanical Society of Edinburgh,* and which was also published in the *Annals & Magazine of Natural History.*[15]

Naturally Spruce had to consider his future employment. At the same time he went through another spell of ill health and in July 1848 he wrote to his friend William Borrer[16] that he had been confined to bed for ten days *"suffering at times the most excruciating pains from the passing of gall stones."* In view of a history of poor health it is perhaps surprising to find that Spruce chose to continue with a career as a plant collector, but with the encouragement of William Borrer and Sir William Hooker he decided to undertake botanical explorations of the Amazon valley.

At that time, A.R. Wallace and H.W. Bates were already collecting insects in the Amazons, and reports were being received from them speaking highly of the climate and of the people.[17] Having made his decision, Spruce was fortunate to enlist the services of George Bentham as his agent in England. This entailed receiving all Spruce's botanical collections, sorting them into sets and sending them on to the various subscribers in Britain, Europe and America. Bentham also undertook to describe many of the new species and genera collected and to keep all the financial accounts. In return for this not inconsiderable work he was to receive the first complete set of plants collected.

With preparations completed, Spruce set off for South America on what was to be one of the most important botanical explorations of the area by one man. In a letter to William Munro, written days before departure[18] he expressed the opinion that he would be absent for just three or four years. In the event, Richard Spruce was to devote the next fifteen years to this task.

ACKNOWLEDGEMENTS

I am grateful to the Borthwick Institute of the University of York, the Linnean Society of London, Manchester City Library, the Royal Botanic Gardens Kew, the Royal Pharmaceutical Society, and the Scarborough and York Public Libraries (North Yorkshire County Council) and Manchester City Library, for access to material in their collections and the assistance of their staff.

NOTES

(1)   R. Spruce, *Notes of a botanist on the Amazon and Andes* (ed. A.R. Wallace), London, 1908, 2 vols.

(2)   R. Spruce, 1867 Letters...to D. Hanbury, Collections of correspondence on materia medica &c., 1851-1908. In the archives of the Royal Pharmaceutical Society. P320 Ms [155].

(3)   R. Spruce, 1844 Letters...to W.Borrer, 1842-1848. In RBG Kew Archives, doc. 32.

(4)   R. Spruce, 1859 Letters...to Bentham, 1842-1848. In RBG Kew Archives, doc. 73.

(5)   R. Spruce, 1864 Letters...to D.Hanbury, Collections of correspondence on materia medica &c., 1851-1908. In the archives of the Royal Pharmaceutical Society. P320 Ms [23].

(6)   R. Spruce, 'Three days on the Yorkshire Moors', *Phytologist* (1841), **1**, 101-104.

(7)   R. Spruce, 'List of mosses &c collected in Wharfedale, Yorkshire', *Phytologist* (1842), **1**, 197.

(8)   R. Spruce, 1843 Letters...to W. Borrer, 1842-1848. In RBG Kew Archives, doc. 3.

(9)   Spruce, op. cit. (1).

(10) R. Spruce, 1867 Letters...to D. Hanbury, Collections of correspondence on materia medica &c., 1851-1908. In the archives of the Royal Pharmaceutical Society. P320 Ms [161].

(11) R. Spruce, 1844 Letters...to W.Borrer, 1842-1848. In RBG Kew Archives, doc. 22.

(12) R. Spruce, 1844 Letters...to W.Borrer, 1842-1848. In RBG Kew Archives, doc. 31.

(13) R. Spruce, 1845 Letters...to W.Borrer, 1842-1848. In RBG Kew Archives, doc. 38.

(14) R. Spruce, 'The Musci and Hepaticae of the Pyrenees', *Transactions of the Botanical Society of Edinburgh* (1850), **3**, 103-216; also published in parts in advance of the above in *Annals and Magazine of Natural History* (1849), **3**, 81-106, 269-293, 358-380, 478-503, **4**, 104-120.

(15) Ibid.

(16)   R. Spruce, 1848 Letters...to W.Borrer, 1842-1848. In RBG Kew Archives, doc. 82.

(17)   A.R. Wallace, *My life: a record of events and opinions*, London, 1905, 2 vols.

(18)   R. Spruce, 1849 Letters...to Munro, 1842-1848. In RBG Kew Archives, doc. 196.

3

Thoughts and observations of Richard Spruce

Winston Spruce

Thoughts and observations of Richard Spruce

Winston Spruce
Wolverhampton, UK

I remember when, as a boy, out walking with my father Lt. Col. Stanley Spruce, he took me to task over my lack of ability to recognise various types of tree. Especially, he said, since our Spruce family shared distant ancestry with the renowned botanist, Richard Spruce.

My father's reference to the botanist stayed in my mind, but it was when I retired from being a schoolmaster, some few years ago, that, looking through a book on *Wild Flowers* by Ron Wilson, I came across a reference to the water lily, *Victoria amazonica* and Richard Spruce's description of its leaves as being like large tea-trays.

Recalling my father's words about Richard Spruce, I began to research my own family history and that of Richard Spruce. Once having taken up his trail, I found that the more I learned of him, the more I wished to know; I began reading and researching all that I could find about Richard Spruce, his own writings and the writings of others.

Genealogical research shows that the ancestors of Richard Spruce, as far back as the early 1700s, were farmers and shepherds living in that part of Yorkshire which lies between the town of Malton and the coastal town of Bridlington. Richard's father was Richard Spruce, schoolmaster at Welburn and Ganthorpe. His grandfather was Richard Spruce, a farmer, of Weaverthorpe and Trensall. His great-grandfather, John Spruce, farmed at Weaverthorpe and had married the daughter of a shepherd. So, we may see that the lives of Richard's forbears were closely linked with nature and the countryside.

Richard's mother, Ann, died and was buried in Terrington in 1829, when Richard was 11½ years of age. Three years later, his father remarried. From this second marriage there were eight daughters.

Richard the botanist had three uncles, William, George and John, born between 1784 and 1789, at which time my own direct Spruce ancestors were thriving in Shropshire. I have found family connections between some Spruces in Yorkshire, Shropshire and Staffordshire.

Following the death of his mother, Richard Spruce began to spend more and more of his free time in making journeys into the local countryside in pursuance of his love of nature and his interest in finding and studying wild plants.

The Ordnance Survey "One inch to the mile" maps, first published early last century, show the Yorkshire countryside as Richard Spruce would have known it. One can see that even within an hour's walk of the Castle Howard villages, he would have found many rewarding locations for plant hunting. From the area close to his home out to the wilder and more remote parts, the crags, the moors and wolds, Richard could gradually accumulate detailed knowledge of the terrain. As a born and bred countryman, having that keen eye and acute powers of observation, he could develop the ability to look over a piece of country and read it like a book. What might be growing in that stony outcrop, that marshy hollow, or on that sunny sheltered bank? Many would be the places known to him. Where to look, and when to look?

So from this, his early hunting ground, he set a trail which, in times before transport and communication as we know them, took him to the Pyrenees, then to the great Rio Amazonas and the Andes, and from Pará right across to Ambato and Guayaquil, with many meanderings and diversions along the way, not to mention dangers and difficulties.

My own impressions of Richard Spruce are based upon his writings, and the writings of those who knew him. I am indebted to the archivists at the Royal Botanic Gardens, Kew for giving me access to Spruce's field journals and other papers.

To read those journals, each entry carrying us back to its place and date of origin, is like being with the botanist on his travels: his beautiful descriptions of the sights, sounds and events of forest and river life. Richard Spruce was interested in, and wrote about all that went on around him during his travels. This, in addition to his careful botanical work. One can be with him making camp at the end of a long day, with whatever has been caught for the pot being prepared and cooked. One can see him carefully spreading out the plants on sheets of drying paper for labelling, packing and dispatch down river to one of his agents; but, above all, one can feel that one has come closer to Richard Spruce. It is then easy to imagine that the gap of time, a century and a half, has narrowed and that we know him.

I see him as a person who, from his very early days, was imbued with a love of learning, having an enquiring mind, being methodical, a perceptive thinker, a careful observer.

From the beginning in Yorkshire, his plant hunting, of necessity, took him to quiet and remote areas. That, combined with his love of study, meant that he became quite accustomed to solitude, but his was an active and lively mind from which came his great written work on the liverworts of the Amazon and Andes.

Those who knew him wrote of his quiet sense of humour, his humility; a man who sought neither fame nor fortune, yet, possessing such great fortitude as was needed to enter little-known fever-ridden country, amongst primitive people. His whole life's goal, the star by which he set his course, was to seek, find and understand the wonders of nature: indeed to be a botanist.

Eminent persons engaged in natural science and associated fields have, at the York Commemorative Conference, paid tribute to Richard Spruce, emphasising the value and importance of his great work as a botanist and the benefits this brought to medicine.

Perhaps it is appropriate here to acknowledge the important work which they themselves have done, and continue to do. The world of nature is never still. It holds secrets yet to be revealed. Present day problems affecting the great rain forests; environmental matters world-wide; the role of plants in the ongoing work to combat human and animal diseases; the need to bring an end to famine. All these problems are being addressed by today's botanists, biologists, environmental scientists, and others. They continue the work done by Richard Spruce and other early explorer botanists.

Then, of course, there is sheer beauty of colour, form and variety in the natural world which man must endeavour to preserve.

4

Tracking Richard Spruce's legacy from George Bentham to Edward Whymper

Joseph Ewan

Tracking Richard Spruce's legacy from George Bentham to Edward Whymper

Joseph Ewan

Missouri Botanical Garden, St. Louis, Missouri, USA

Introduction

For the United States 1849 was the year "the world rushed in" and the goldfields magnetized men. For the field biology of South America it was July 12, 1849, when the *Britannia*, a brig of 217 tons, docked at Pará at the mouth of the Amazon, with Richard Spruce aboard. Every man faces a race with oblivion: Richard Spruce is not a familiar name even among fellow botanists today. How did this happen? Alfred Russel Wallace, "Darwin's moon", whom Spruce knew in Brazil, and Henry Walter Bates, also in Spruce's company, are known to many freshman biology students. If only Spruce had codified one of his many field observations: Wallace is remembered for Wallace's Line; Bates, for mimicry. Fame is fickle.

Spruce "was more than a botanist". He was an explorer, an ethnobotanist among the Amerindians of South America, an acute observer of the ecosystem, as we call it today, of physiography in the tradition of Boussingault,[1] whose work on the Andes he owned. Like Clements Markham he was fascinated by the history of peoples. In the world of linguistics he noted vocabularies of plant names used by the natives. In plant taxonomy he focused on *Hevea* among flowering plants. Spruce made the critical collections of seven species of *Hevea* described by three botanists.[2] By collecting over 7,000 specimens of plants, perceptively identified by George Bentham and distributed to herbaria around the world, Spruce made known the South American flora. Today, 43 herbaria hold priceless vestiges of species preserved by Spruce that in some instances have been exterminated in the wild, or teeter on the edge of extinction.

George Bentham, 1800–1884

With Spruce's specimens before them, George Bentham and Joseph Dalton Hooker, working across the table, were able to sharpen their generic concepts as they codified plant groups comprising the world's flora in the writing of their monumental *Genera plantarum*.[3] Illustrative of Spruce's engagement with the ecology of the Amazon was that curious euphorb, *Phyllanthus fluitans*, which recalls the floating *Salvinia*. I cannot forget the day at the Smithsonian when a specimen arrived. As customary, the unknown plant was passed around the herbarium – we need not recite all those who failed the question – but it was Conrad Morton, the pteridologist, who identified the anomalous euphorb! Spruce wrote to Bentham on November 7, 1851, regarding the puzzling waterplant;[4] Hooker[5] published Bentham's letter in part:

> *"By the bye, our little Phyllanthus fluitans was in abundance* [at Manaquiry]: *are you sure that the embryo of this is dicotyledonous? – there is a remarkable analogy, to say the least, with Hydrocharis"*.

Alfred Russel Wallace, 1823–1913

Another legacy of Spruce is his journal of travels in South America, made known in print through the devotion of Alfred Russel Wallace, published at Wallace's expense and at some loss.[6] It is the primer for every would-be botanical explorer seeking a Baedeker. Paul Cutright[7] wrote that

> *"Wallace did remarkably well in editing Spruce's notes; but we wonder what Spruce would have told the reader if he himself could have written the complete story of his fifteen years under the tropical sun in America".*

However, Schultes[8] offered another thought: "Spruce never could have written a book about Spruce – he was too self-effacing and humble". John Murray was evidently in correspondence with Spruce on a projected book.[9]

Spruce and Wallace took a particular interest in palms. With what the Cambridge botanist Corner called "his usual perspicacity", Spruce recognized fourteen species of *Geonoma* whereas Martius had described two. Wallace named three species of *Bactris*; Spruce knew fifteen. Spruce related the palms of the Amazon to the geological history of the basin.[10] But all the while Spruce exuded modesty. He wrote that Martius gave us the *"noblest monograph of any family of plants which has ever issued from the press"*, adding that he was the *"most eminent botanist who had ever visited South America"*.[11] Though his specimens were less than ideal, Spruce collected critical palm fruits and foliage that still survive at Kew.

In writing his *Palmae amazonicae* Spruce examined the herbarium records of George Gardner, William Purdie, and Jules Linden to complement his own descriptions. In his 118-page recension of palms Spruce described what he called the "alternation of function", that is the timing of separate and pistillate flowering on the same individual year by year. He had revisited a plot of *Geonoma* palms which had "borne young fruits" in May, 1852 and made a further visit the following year. He wrote:

> *"I saw, to my astonishment, the very same plants all bearing male flowers alone! But the mystery disappeared when, on examination, I made out that male and female spadices must have alternated all the way up the stem".*

Although Professor Corner did not mention Spruce's thesis on the "alternation of function" in his comprehensive *Natural History of Palms*, he reported on the separate flowering periods on the same individual in the African oil palm (*Elaeis guineenis*). Corner[12] remarks that the cause of this differentiation is unknown, but it may suggest the evolution of the dioecious habit.

James Orton, 1830–1877

The American naturalist-explorer James Orton pronounced Richard Spruce to be *"the most accomplished botanist on the Amazons since von Martius left it"*. He noticed Spruce's "alternation of function" in his third edition of his *Andes and Amazon*.[13] Orton made three expeditions to South America, the first crossing from Guayaquil to Pará in 1867, the second in the opposite direction, and the third originating on the coast of Peru, which ended with Orton's death on Lake Titicaca.[14] His *Andes and Amazon*, dedicated to Charles Darwin, went through three editions. He devoted a chapter to the palms and acknowledged Spruce's "eleven years of research" which resulted in what Orton called *"a well-nigh inaccessible memoir"* on palms, published in the Linnean Society's *Proceedings*. One wonders where Professor Orton, then corresponding from Vassar College with specialists from Montreal to Washington, finally found a copy of Spruce's memoir?

William Jameson, 1796–1873

William Jameson, a native of Edinburgh, arrived in Callao as a ship's surgeon in 1820.[15] Later he took up residence in Quito for his health and lived there for 44 years, teaching botany and chemistry at the University. He is remembered for his pioneer flora of Ecuador,[16] modestly titled a *Synopsis*, that included an unfinished third volume. We do not know when Jameson first corresponded with Spruce but the introduction was probably made by W.J. Hooker. Jameson also corresponded with John Lindley and later with H.G. Reichenbach concerning orchids, and with John Gould for whom he collected hummingbirds. Rounds[17] mentioned, without details, that Spruce had eleven customers for bird skins, perhaps arranged by Jameson? We know that Jameson wrote to Hooker at Kew in 1858: *"it is now several months since Mr. Spruce was in the interior* [of Ecuador] – *At present he is in Ambato, only two days journey from* [Quito]".[18] Jameson's letters to Hooker extend from 1850 to 1859 and in that letter of 1858 he had remarked that he had not yet met Spruce but had letters from him, adding *"I shall most cheerfully accompany him on his explorations to the more interesting localities in this neighborhood"*. In 1859 Jameson wrote to Hooker that *"Spruce is at present toward the base of Tungurara and Condopasto and has lately made some interesting explorations of these mountains"*.[19] An overlooked association is that James Orton considered completing Jameson's *Synopsis* of the flora of Ecuador, a plan terminated by Orton's death in 1877. Perhaps the fragment of Jameson's projected third volume, now at the Smithsonian, was delivered by James Orton in his friendly association with Smithsonian's Spencer Fullerton Baird?

Friedrich Hassaurek, 1831–1885

Another Quito connection with Spruce was Friedrich Hassaurek. Born in Vienna in 1831, Hassaurek was a "forty-eighter" who arrived in Cincinnati in 1849, became prominent in journalism and politics and supported Abraham Lincoln's presidency. It was from this advocacy that he was appointed United States Minister in Ecuador. He cited Spruce seven times in his engaging and informative *Four Years among Spanish Americans*, published in New York.[20] He was acquainted with Ecuador's history and geography and made botanical references, for example to the genus *Espeletia*. Though we have no record of them meeting, it seems probable that they did, from the quotations of Spruce's writings by Hassaurek.[21] Hassaurek[22], who died in Paris in 1885, considered Spruce *"a botanist of high standing in the scientific world"*.

Heinrich Gustav Reichenbach, 1824–1889

Born in Leipzig, the son of the author of the *Icones Florae Germanicae et Helveticae*,[23] Reichenbach developed a great interest in orchids in his youth, while collaborating with his father in the publication of the *Icones*. Merle Reinikka provides a short account of Reichenbach and his relations with the English botanist John Lindley. Reinikka[24] says that Reichenbach's letters were often *"tinged with wit and sarcasm"* and perhaps they may contain references to Spruce. Reichenbach published an enumeration of Spruce's plant collections arranged by geographic regions in the journal *Botanische Zeitung*[25] in 1873. Spruce was awarded an honorary degree of Doctor of Philosophy by the Imperial German Academy – Academia Caesarea Leopoldina-Carolina – in 1865, perhaps through the support of this Hamburg University professor. John Lindley, the correspondent of Spruce and friend of Kew, was Reichenbach's ally in the classification of orchids. Among orchidologists Reichenbach left a memory of "impetuous temper" around his "consuming passion" for orchids. For Spruce it was a lifetime enthusiasm for all plants. After Lindley's death in 1865 Reichenbach became the "orchid king". At least three times Lindley had commemorated Spruce on the basis of his collections of Brazilian orchids. The details of the Reichenbach-Spruce friendship remain to be disclosed.

Clements Robert Markham, 1830–1896

Sir Clements Markham, traveller, geographer and author, was born at Stillingfleet near York in 1830. After serving with the Franklin search expedition in the Arctic, Markham retired from the navy and devoted the years 1852–1854 to travels in Peru. Later he contributed introductions to twenty publications for the Hakluyt Society, of which he was secretary from 1858 to 1887 and president in 1890. Markham is widely noticed for engineering the successful introduction of *Cinchona* to India; Albert Markham[26] wrote: *"with the zeal and untiring efforts of* [Mr. Spruce to whom] *a large share of the success of the enterprise was due"*. Markham presented Spruce with a copy of his book *Expeditions into the Valley of the Amazons, 1539, 1540, 1639*.[27] Spruce neatly annotated this copy with his marginalia; for example, in the chapter on Gonzalo Pizarro he wrote, *"if it were impossible to ascend the river in Gonzalo's canoes, how much more so in Orellano's brigantine"*. I purchased this association copy at a sale of selected books from Spruce's library; it is now at the Missouri Botanical Garden.

Edward Whymper, 1840–1911

Edward Whymper, the British alpinist, explorer and artist, succeeded in 1865 in reaching the summit of the Matterhorn after six attempts. He then organized an expedition to Ecuador designed to study the biology of high altitudes, including mountain-sickness. Chimborazo was then believed to be the highest mountain in the Americas. When Whymper first saw Chimborazo on December 21st, 1879, he was surprised to find it consisted of two summits, later finding the eastern to be the higher one. Humboldt and Boussingault had mentioned only a single summit. Whymper calculated the eastern summit to be 20,608 feet, an altitude never before reached. He found twenty plant families growing above 15,000 feet, according to determinations made by William Carruthers of the British Museum (Natural History). He also collected moths and butterflies, and found earthworms up to 15,871 feet elevation. Whymper reported glaciers that had been overlooked by Humboldt and Boussingault and named a glacier after Spruce, whose field studies he knew. Whymper's *Travels amongst the great Andes of the Equator*, first published in London in 1891[28] is a stunning work, particularly the handsome subscribers' edition, London, 1892. William Trelease, botanist and avid bookman at the Missouri Botanical Garden, who had acquired the 1896 New York edition of Whymper's *Travels*, was interested in his plant records. Perhaps he knew the Whymper plant collections recorded in the two pages of the unpublished Kew Plant Lists. In any event Trelease must have directed J.B.S. Norton, a student at Missouri Botanical Garden, to prepare a $6^1/_2$-page handwritten index of "everything concerning botany in the *Travels*". And so Whymper, who lives on in his benchmark *Travels* on his exploration of Chimborazo and Cotopaxi, profited from the experience of Spruce, and Spruce lives on in the name of the glacier on Chimborazo, Glacier de Spruce![29]

Isaiah Bowman, 1878–1950, and Teodoro Wolf, 1841–1924

Spruce, the geographer, studied the *gapo* and the *terra firma* or, as he called it, the *Urwald*. Soils interested Spruce, especially on the opposite side of the Amazon at Manaus where the Rio Negro joins the Solimoes. Isaiah Bowman was attracted by a dry pocket in the Huallaga basin described by Spruce. Bowman called it a *"deep semi-arid pocket with only a patchy forest"*. Based on three expeditions to South America in 1907, 1911, and 1913, Bowman[30] reported in his *Andes of Southern Peru* on the topography, climatology, and in brief on the vegetation. After serving as Director of the American Geographical Society and fostering the international Millionth Map of Hispanic America, he left foundation works that complement the writings of Teodoro Wolf, another geographer who noticed Spruce. His *Geografica y Geologia del Ecuador*[31] summarized the vegetation of Ecuador in the context of her physiography, based on careful cartography.

Christopher Sandeman, 1882–1951

"An heroic botanical odyssey in South America [began] *when Spruce arrived in Para in 1849"*. So wrote Christopher Sandeman, of Portuguese ancestry and a resident of Spain for many years, who followed in the tracks of Spruce in the Andes. A century after Spruce arrived, Sandeman[32] wrote his sensitive appreciation of Spruce. He often revered Spruce's modesty and keen observation, and

his sympathies with the field botanist are documented by his herbarium records at Kew. Here is his tribute to Spruce:

> *"Never perhaps was botanist or plant collector more unsparing of himself, more singly devoted to his life-work, more philosophical in his acceptance of appalling discomforts, more heroic in his endurance of physical pain as a normal way of life. Love for Nature and curiosity about her processes burnt in him like a flame, and Richard Spruce had not even a bowing acquaintance with self pity".*

Sandeman's photographs in his *Wanderer in Inca Land*[33] carry artistic merit of their own.

Victor Wolfgang Friedrich Heinrich von Hagen, 1908–

Sandeman wrote a testament of Spruce for his British readers; Victor von Hagen, for the Americans. A native of St. Louis, Missouri, von Hagen first collected termites, then with the appeal of anthropology collected the lore of the Mayas, and later of the Incas. He wrote a shelf of books, often with facile descriptions and generally illustrated with photographs of unusual quality. In 1935 he led an expedition to the Galapagos to mark the centenary of Charles Darwin's visit. Hickman[34] wrote: *"von Hagen's fertile mind conceived the idea of a scientific research station"*, though it was twenty-five years before the Galapagos research station was founded. Two of von Hagen's travel books, *Off with their heads* and *Ecuador the unknown*[35] acknowledge Spruce's philosophy. *"The green world awaits a man who will carry on the work of Richard Spruce"*, wrote von Hagen,[36] *"one who will be deeply stirred by those green mansions"*, who, he adds, is *"not a mere collector"*. One of the most needed works, he said, is an overall view of the plant geography of South America: *"The list of things yet to be done is almost endless"*.

Conclusion

With George Bentham as overseas herbarium agent to Richard Spruce; Wallace as his devoted editor-reporter; William Jameson, his way-station encourager in Quito; with James Orton, Spruce's North American advocate as early as 1870, Spruce's supporters made up a valiant company. Then there was commentator-diplomat Hassaurek, who knew Spruce in Quito and often quoted from his writings; orchidologist Reichenbach prized his *exsiccatae*; alpinist Whymper, encouraged by Spruce's findings in the Andes, botanized on the summits and named a glacier to commemorate him; geographers Teodoro Wolf and Isaiah Bowman left tracks of Spruce in their classic works on the Andes. Then there was worshipful Christopher Sandeman who knew the cliff faces of the Andes and through his prose transported readers from their cottages to Inca-land. At the same time Victor von Hagen in book after book revived Richard Spruce for the armchair explorer. Finally, Spruce's Boswell, Richard Evans Schultes[37] has insisted that *"the botanist is wrong in claiming Spruce wholly for botany ...* [he remains] *the all-round scientist and man of culture ... sorely missed* [today] *and so urgently needed"*. Richard Spruce has won his race with oblivion.

ACKNOWLEDGEMENTS

My appreciation to colleagues here at the Garden for favours, and importantly to Richard Schultes who has sent me copies of his essential writings of Richard Spruce over the years. John Hoover of the St. Louis Mercantile Library has searched for archival records of Victor von Hagen preserved in his native city. Again my wife Nesta has suggested details, smoothed the rough places, and better posted Spruce's track.

NOTES

(1)   J.B.J.D. Boussingault, *Viajes científicos a los Andes ecuatoriales*, Paris, 1849.

(2)   R.E. Schultes, 'A brief taxonomic view of the genus *Hevea*', Kuala Lumpur, 1990. (*Malaysian Rubber Research and Development Board* Monograph 14).

(3)   B.D. Jackson, *George Bentham*, London, 1906.

(4)   R. Spruce, *Notes of a botanist on the Amazon and Andes*, (ed. A.R. Wallace), London, 1908, vol. 1, 115, 230. [Reprinted edition, with a new foreword by R.E. Schultes, New York, 1970.]

(5)   R. Spruce to G. Bentham, 7 November 1851, in 'Botanical information: intelligence of Mr Spruce in a letter to G. Bentham, Esq.', *Hooker's Journal of Botany and Kew Gardens Miscellany* (1852), **4**, 280.

(6)   C. Sandeman, ed., *No music in particular*, London, 1943, 276.

(7)   P.R. Cutright, *The great naturalists explore South America*, New York, 1940.

(8)   R.E. Schultes, 'Preface' in Spruce, op. cit. (4) [1970 reprint], v.

(9)   G. Stabler, 'Obituary notice of Richard Spruce Ph.D.', *Transactions and Proceedings of the Botanical Society Edinburgh* (1894), **20**, 106.

(10) E.J.H. Corner, *Natural history of palms*, London, 1966, 36.

(11) R. Spruce, 'Palmae amazonicae, sive enumeratio palmarum in itinere sue per regiones Americae aequatoriales lectarum', *Journal of the Linnean Society Botany* (1869), **11**, 65-183.

(12) Corner, op. cit. (10), 136.

(13) J. Orton, *Andes and Amazon*, ed. 3, New York, 1876, 537.

(14) J. Ewan, 'Five American naturalists on the Andes and Amazon: a study in personality', *Archives of Natural History* (1989), **16**, 2.

(15) I. Anderson-Henry, 'Biographical notice of Professor Jameson of Quito', *Transactions and Proceedings of the Botanical Society Edinburgh* (1876) **12**, 22.

(16) W. Jameson, *Synopsis plantarum aequatoriensium*, Quito, 1865, 2 vols.

(17) R.S. Rounds, *Men and birds in South America, 1492-1900*, Fort Bragg, Calif., 1990, 70-71. ["Robert Spruce" lapsus typographicus.]

(18)   W. Jameson, Letter to Sir William Hooker, from Quito, 24 March. In RBG Kew Archives. Directors' Correspondence, Vol. LXXI, doc. 178. 1858.

(19)   W. Jameson, Letter to Sir William Hooker from Quito, 19 January. In RBG Kew Archives. Directors' Correspondence, Vol. LXXI, doc. 179. 1859.

(20)   F. Hassaurek, *Four years among Spanish Americans*, New York, 1867.

(21)   Ibid., 78-82.

(22)   Ibid., 15.

(23)   H.G.L. Reichenbach, *Icones florae germanicae et helveticae*, Leipzig, 1834-1914, 25 vols.

(24)   M.A. Reinikka, *A history of the orchid*, Coral Gables, Fla, 1972, 215-218.

(25)   H.G. Reichenbach, 'Zum geographischen Verständniss der amerikanischen Reisepflanzen des Herrn Dr. Spruce', *Botanische Zeitung* (1873), **31**, 28-29.

(26)   A.H. Markham, *Life of Sir Clements R. Markham*, London, 1917, 173.

(27)   C.R. Markham, *Expeditions into the Valley of the Amazons, 1539, 1540, 1639*, London, 1859.

(28)   E. Whymper, *Travels amongst the great Andes of the equator*, London, 1891.

(29)   Ibid.

(30)   I. Bowman, *Andes of southern Peru*, New York, 1916, 153.

(31)   T. Wolf, *Geografica y geologia del Ecuador*, Lepizig, 1892.

(32)   C. Sandeman, 'Richard Spruce, portrait of a great Englishman', *Journal of the Royal Horticultural Society* (1949), **74**, 531-544.

(33)   C. Sandeman, *A wanderer in Inca-Land*. London, 1948.

(34)   J. Hickman, *The enchanted islands. The Galapagos discovere*d, Oswestry, Shropshire, 1985, 115.

(35)   V.W.F.H. von Hagen, *Off with their heads*, New York, 1937; V.W.F.H. von Hagen, *Ecuador the Unknown*, New York, 1939.

(36)   V.W.F.H. von Hagen, *Off with their heads*, New York, 1937, xix.

(37)   R.E. Schultes, 'Preface' in Spruce, op. cit. (4) [1970 reprint], ix.

FURTHER REFERENCES CONSULTED

J. Ewan, 'Through the jungle of Amazon travel narratives of naturalists', *Archives of Natural History* (1992), **19**, 185-207.

J. Orton, *Andes and Amazon*. New York, 1870.

C. Sandeman, *Thyme and bergamot*. London, 1947.

R.E. Schultes, Richard Spruce still lives. *Northern Gardener* (1953), 7, 20-27, 55-61, 87-93, 121-125.

V.W.F.H. von Hagen, *South America called them. Explorations of the great naturalists: Condamine, Humboldt, Darwin, Spruce*. New York, 1945.

V.W.F.H. von Hagen, *Ecuador and the Galapagos Islands*. Norman, Okla., 1949.

5

With Humboldt, Wallace and Spruce at San Carlos de Río Negro

Duncan M. Porter

With Humboldt, Wallace and Spruce at San Carlos de Rio Negro

Duncan M. Porter

*Department of Biology, Virginia Polytechnic Institute and State University,
Blacksburg, Virginia, USA*

Introduction

The village of San Carlos de Rio Negro is at latitude 1°56´N, longitude 67°03´W in Venezuela's Territorio de Amazonas. About 15 kilometres north of San Carlos, the Rio Guainia and the Rio Casiquiare merge to form the Rio Negro; to the south, the Rio Negro eventually joins the Rio Amazonas, the Amazon.

San Carlos lies very near the borders of Venezuela, Colombia, and Brazil, and has long played a strategic role. It was founded as a military post in 1759. It soon had a Franciscan mission, in order to persuade (or indoctrinate) the Indians of the benefits of the Spanish empire, rather than that of Portugal across the ill-defined border to the south.

I was in San Carlos in 1975 as an observer on a team from the US National Science Foundation, making a site visit to a project we were funding that was studying nutrient cycling in the rain forest.[1] Despite its remoteness, San Carlos was chosen as a research site because there was a Venezuelan Air Force landing strip nearby. Whereas it might take weeks to get to San Carlos overland from north to south, we flew there from Caracas in about eight hours. It took Alexander von Humboldt and the botanist Aimé Bonpland three months to make the trip in 1800, although admittedly they stopped to observe and collect along the way.

Humboldt

On 5 June 1799, the 29-year old Prussian Friedrich Heinrich Alexander, Baron von Humboldt and the 25-year old Frenchman Aimé Jacques Goujaud Bonpland set sail from Spain on a scientific expedition to South America. It was their object, Humboldt later wrote,

> *"to make known the countries I had visited; and to collect such facts as are fitted to elucidate a science of which we as yet possess scarcely the outline, and which has been vaguely denominated Natural History of the World, Theory of the Earth, or Physical Geography. The last of these two objects seemed to me the most important."*[2]

Humboldt and Bonpland left with the blessings of King Carlos IV of Spain, who gave them permission to collect and observe in his New World possessions. They were to sail to Havana, but sickness on the ship (*"a malignant fever"*)[3] caused the captain to put into the port of Cumaná in Nueva Andalucia, now in Venezuela, where they landed on 16 July 1799. After collecting in the northern part of the country for almost six months, they left Caracas on 7 February 1800, bound for the Rio Orinoco in the far south. The most important objective of this journey was

> "to determine astronomically the course of that arm of the Orinoco which falls into the Rio Negro, and of which the existence has been alternately proved and denied during half a century."[4]

Should the Rio Casiquiare prove navigable, it might provide a way to increase commence between the Spanish and Portuguese possessions and open the area up to increased exploitation. Humboldt later wrote:

> "In proportion as the activity of commerce increases in these countries traversed by immense rivers, the towns situated at their confluence will necessarily become bustling ports, depots of merchandise, and centre points of civilisation."[5]

Humboldt and his party reached the Orinoco and paddled up it and the Rio Atabapo, then travelled overland to the Rio Guiania, their boat being portaged by 23 hired Indians. They then floated south to San Carlos, arriving on 7 May 1800. Humboldt's *Personal Narrative*, originally published in French in 23 volumes (1805-1834), and first translated into English in 1814-1829 (7 volumes), has a great deal of information on the biology, geography, and geology of the region, but has little on San Carlos, where they spent only three nights:

> "We lodged at San Carlos with the commander of the fort, a lieutenant of militia. From a gallery in the upper part of the house we enjoyed a delightful view of three islands of great length, and covered with thick vegetation. The river runs in a straight line from north to south, as if its bed had been dug by the hand of man. The sky being constantly cloudy gives these countries a solemn and gloomy character. We found in the village a few juvia-trees, which furnish the triangular nuts called in Europe the almonds of the Amazon, or Brazil-nuts. We have made it known by the name of Bertholletia excelsa. The trees attain after eight years' growth the height of thirty feet."[6]

Humboldt also wrote of the difficulties in collecting plants near San Carlos:

> "All these trees (with the exception of our new genus Retiniphyllum [a genus of Rubiaceae; there are four species near San Carlos]) were more than one hundred or one hundred and ten feet high. As their trunks throw out branches only towards the summit, we had some trouble in procuring both leaves and flowers. The latter were frequently strewed upon the ground at the foot of the trees; but, the plants of different families being grouped together in these forests, and every tree being covered with lianas, we could not, with any degree of confidence, rely on the

authority of the natives, when they assured us that a flower belonged to such or such a tree. Amid these riches of nature he[r]borizations caused us more chagrin than satisfaction. What we could gather appeared to us of little interest, compared to what we could not reach. It rained unceasingly during several months, and M. Bonpland lost the greater part of the specimens which he had been compelled to dry by artificial heat. Our Indians distinguished the leaves better than the corollae or the fruit. Occupied in seeking timber for canoes, they are inattentive to flowers. "All those great trees bear neither flowers nor fruits," they repeated unceasingly. Like the botanists of antiquity, they denied what they had not taken the trouble to observe. They were tired with our questions, and exhausted our patience in return."[7]

Humboldt refers to difficulty crossing the border into Brazil due to political differences between the Spanish and the Portuguese at the time. However, he does not relate that he actually crossed into Brazil, where he was treated as a spy. He was arrested, and his collections, notebooks, drawings, and instruments were confiscated. The Portuguese authorities had been given orders to bar him from their territory. The official *Gazeta da Colônia* in Pará on 2 July 1800 announced that:

"... a certain Baron von Humboldt, from Berlin, has been travelling through the interior of America, making astronomic observations in order to rectify certain errors in the existing maps, and collecting plants ... Under this pretext this stranger may hide plans for the propagation of new ideas and new religious principles among the loyal subjects of this domain. His excellency [the Governor of Pará] *should investigate the case ...; otherwise, it would be extremely dangerous to the political interests of the Portuguese crown, if this were the case ..."* [8]

By this time it was too late. After much discussion, Humboldt was released, his possessions were returned to him, and he returned to Spanish territory. Humboldt and Bonpland left San Carlos having tarried there only a few days. They travelled north up the Casiquiare to the Orinoco, proving for the first time that it could be done, and paddle out of our story.

Wallace

When Alfred Russel Wallace was teaching in Leicester in 1844-45, he met Henry Walter Bates, who was working as an apprentice in a warehouse. Bates was an ardent entomologist who inspired Wallace to take an interest in insects; their early correspondence is full of discussions of these animals. A few years later their close friendship and common interests led Wallace to suggest that they make a collecting trip in the tropics. Wallace, aged 25, and Bates, then 22, left England for Brazil on 25 April 1848. Wallace later wrote in his book on his adventures, *A Narrative of Travels on the Amazon and Rio Negro*, published in 1853, that:

"An earnest desire to visit a tropical country, to behold the luxuriance of animal and vegetable life said to exist there, and to see with my own eyes all those wonders which I had so much delighted to read of in the narratives of travellers, were the motives that induced me to break through the trammels of business and the ties of home, and start for 'Some far land where endless summer reigns.'

My attention was directed to Pará [now Belém] and the Amazon by Mr Edwards's little book, "A Voyage up the Amazon,"[(9)] *and I decided upon going there, both on account of its easiness of access and the little that was known of it compared with most other parts of South America.*

I proposed to pay my expenses by making collections in Natural History, and I have been enabled to do so; and the pleasures I have found in the contemplation of the strange and beautiful objects continually met with, and the deep interest arising from the study in their native wilds of the varied races of mankind, have been such as to determine my continuing in the pursuit I have entered upon, and to cause me to look forward with pleasure to again visiting the wild and luxuriant scenery and the sparkling life of the tropics."[(10)]

Ten years later, in his own book on the visit, Bates[(11)] wrote that he and Wallace went to South America:

"to make for ourselves a collection of objects, dispose of duplicates in London to pay expenses, and gather facts, as Mr Wallace expressed in one of his letters, towards solving the problem of the origin of species, a subject on which we have conversed and corresponded much together".

Wallace and Bates arrived in Pará on 28 May 1848. Before leaving England, they were commissioned by Sir William Hooker to collect plants for the Royal Botanic Gardens, Kew, and the British Museum agreed to buy any rare insects that they might find. During their first two years in Brazil, Wallace and Bates collected along the Amazon and its tributaries, twice meeting up with Richard Spruce. They spent much of 1850 in and around Barra do Rio Negro. Bates had left to explore the Solimões, the main stream of the Amazon, on 26 March. They did not meet again for over a decade.

Travelling up the Rio Negro and its tributaries, collecting on the way, Wallace arrived in San Carlos from the south on 4 February 1851. He was now in Venezuela, a country formed in 1829 from some of the remnants of the Spanish New World empire. Wallace later wrote:

"At length, on the afternoon of the 4th of February, we arrived at São Carlos, the principal Venezuelan village on the Rio Negro. This was the furthest point reached by Humboldt from an opposite direction, and I was therefore now entering upon ground gone over fifty years before by that illustrious traveller. At the landing-place I was agreeably surprised to see a young Portuguese I had met at Guia, and as he was going up the river to Tomo in a day or two, I agreed to wait and take him with me. I went

with him to the house of the Commissario, got introduced, and com-
menced my acquaintance with the Spanish language. I was civilly
received, and found myself in the midst of a party of loosely-dressed gen-
tlemen, holding a conversation on things in general. I found some
difficulty in making out anything, both from the peculiarity of accent and
the number of new words constantly recurring; for though Spanish is
very similar to Portuguese in the verbs, pronouns, and adjectives, the
nouns are mostly different, and the accent and pronunciation peculiar.

We took our meals at the Commissario's table, and with every meal
had coffee, which custom I rather liked. The next day I walked into the
forest along the road to Soláno, a village on the Cassiquiare. I found a
dry, sandy soil, but with very few insects. The village of São Carlos is laid
out with a large square, and parallel streets. The principal house, called
the Convento (monastery), where the priests used to reside, is now occu-
pied by the Commissario. The square is kept clean, the houses
whitewashed, and altogether the village is much neater than those of
Brazil. Every morning the bell rings for matins, and the young girls and
boys assemble in the church and sing a few hymns; the same takes place
in the evening; and on Sundays the church is always opened, and service
performed by the Commissario and the Indians."[12]

After spending several days in San Carlos, Wallace travelled as far north as Javita on the Rio Temi, a branch of the Rio Atabapo, up which Humboldt and Bonpland had paddled 51 years before. Here, Wallace wrote of the palms, which attracted his attention throughout his travels. Later, he published *Palm trees of the Amazon*,[13] which described 14 new species. Three of his new species still grow around San Carlos. Wallace left Javita for Brazil on 31 March 1851, paddling south by San Carlos in early April, never to return. He encountered Spruce in Barra on September 1851, and stayed with him until leaving for the Brazilian part of the Rio Negro at the end of October. The two had a final meeting at São Gabriel in November.

Wallace's descriptions of San Carlos and the surrounding area lack the detail provided by Humboldt in his *Personal narrative*. This is not surprising, however, considering what happened on his voyage home. Wallace left Barra on 10 June 1852 with his extensive collections made over the past two years and sailed down the Amazon to Pará, arriving on 2 July. Ten days later, Wallace quit Pará and Brazil for the last time, sailing for England. Twenty-five days later, on 6 August 1852, Wallace was reading in his cabin after breakfast when he had a visitor. It was the ship's captain, who said to him, *"I'm afraid the ship's on fire; come and see what you think of it."*[14] The hold was full of rubber, balsam, and other plant products, which were burning fiercely. It soon became apparent that the fire could not be put out, and that the ship must be abandoned. Wallace wrote:

"I went down into the cabin, now suffocatingly hot and full of smoke, to
see what was worth saving. I got my watch and a small tin box contain-
ing some shirts and a couple of old note-books, with some drawings of

plants and animals, and scrambled up with them on deck. Many clothes and a large portfolio of drawings and sketches remained in my berth [this did not include his collections in the hold]; *but I did not care to venture down again, and in fact felt a kind of apathy about saving everything, that I can now hardly account for."*[15]

Wallace *"rescued his watch, sextant, notes and sketches of palm trees, some notes for a map of the Rio Negro and Uaupés rivers, 203 sketches of rare Amazonian fish, and his journals of the first and last portion of his trip."*[16] Everything else was lost; all his carefully and sometimes painfully collected living, pinned, or stuffed animals, the dried plants, and the other notes and sketches were gone. His trials were not yet over; those who had been on the ship floated in the long-boat for nine days and suffered much from thirst before being rescued. On his return to England, Wallace could only use the materials he had rescued to prepare his *Narrative of travels*[17] and his *Palm trees of the Amazon*.[18] Think what he might have done if everything had survived.

Spruce

Richard Spruce was 31 years old when he sailed from England to Brazil on 7 June 1849. Spruce arrived in Pará on 12 July, accompanied by his assistant Robert King and by Herbert Wallace, on his way to join his elder brother Alfred in exploring the Amazon valley. Many years later, Alfred Wallace prepared Spruce's *Notes of a botanist on the Amazon and Andes*[19] for publication. According to Wallace's biographical introduction:

"His decision having been taken, his whole time must have been fully occupied till the date of sailing ... Letters from Sir W[illiam] Hooker in October and November 1848 show that this journey was under discussion, and that by December it was finally decided upon. A letter to [his friend] Mr G[eorge] Stabler shows that Spruce came to Kew in April 1849, and spent about two months there. During this time Mr. Bentham agreed to receive all his botanical collections, name the already described species, sort them into sets under their several genera, and send them to their various subscribers in Great Britain, as well as in different parts of Europe. He also undertook to describe the more interesting new species and genera, and to collect the subscriptions and keep all accounts, in return for which invaluable services he was to receive the first (complete) set of the plants collected.

Later letters show that only eleven subscribers were obtained at first; however, after the early collections arrived and were reported on by Sir W[illiam] Hooker in the Journal of Botany, and by so great a botanist as Mr Bentham, subscribers were at once found for twenty sets, which, a few years later, when the great novelty of the collections and their admirable condition as specimens became more widely known, increased to over thirty."[20]

On his arrival in Brazil, Spruce collected for three months in the vicinity of Pará:

"The beginning of the dry season is a sort of spring in the Amazon valley. As the rains abate and the rivers subside, the trees begin to flower, first those of the gapó or inundated river-margins, then those of the terra firme or dry land. Some trees flower ere the old leaves fall off, others along with the young leaves. In either case the trees are never denuded of leaves, except in a few cases of extreme rarity, the old leaves hanging on until the young ones are developed, exactly as in evergreens at home. A few months later and it is the height of summer; flowers are scarce, and most trees are ripening their fruits and seeds. Both flowers and fruits of the real forest trees were for a long time "sour grapes" to me. Like Humboldt, I was at first disappointed in not finding agile and willing Indians ever ready to run like cats or monkeys up the trees for me, and in seeing how futile must be the attempt to reach with hooked knives fastened to poles flowers which grew at a height of a hundred or more feet, on trees whose smooth trunks (far too thick to be "swarmed") rose to 50 or 60 feet before putting forth a branch. At length the conviction was forced upon me that the best and sometimes the only way to obtain the flowers or fruits was to cut down the tree; but it was long before I could overcome a feeling of compunction at having to destroy a magnificent tree, perhaps centuries old, merely for the sake of gathering its flowers. By little and little I began to comprehend that in a forest which is practically unlimited - near three millions of square miles clad with trees and little else but trees - where even the very weeds are mostly trees, and where the natives themselves think no more of destroying the noblest trees, when they stand in their way, than we the vilest weeds, a single tree cut down makes no greater a gap, and is no more missed, than when one pulls up a stalk of groundsel or a poppy in an English cornfield. I considered further that my specimens would be stored in the principal public and private museums in the world, and would serve to identify any particular tree with its products, as well as for studying the peculiarities of its structure. In time, I reconciled myself to the commission of an act whose apparent vandalism was, or seemed to be, counterbalanced by its necessity and utility. In the same way I suppose a zoologist stifles his qualms of conscience at killing a noble bird or quadruped merely for the sake of its skin and bones. I know not whether Alexanders and Napoleons make use of any such process of reasoning to justify to themselves the waste of human life entailed by their victories; but if the bodies of the slain at Arbela or Austerlitz could all have been collected and preserved - stuffed and set up in attitudes of mortal agony - under glass cases in some vast museum, what instructive specimens they would have been of the fruits of war!"[21]

Spruce advanced to Santarém in October 1849, where he first met Wallace:

"At Santarem I had the pleasure of meeting Mr A.R. Wallace, of becoming acquainted with the paths across the camp under his guidance,

and of his animated and thoughtful conversation in the evenings; although, after a hard day's work, we both of us found it difficult to keep our eyes open after 8 o'clock, for it was not until I had been some time longer in the country that I got into the way of taking a short siesta in the heat of the day, which enabled me to enjoy the evenings more."[22]

By December 1850, Spruce reached Barra do Rio Negro, which he made his headquarters for the next eleven months. On 14 November 1851, he began making his way up the Rio Negro. He methodically collected along this river and the Rio Uaupés, reaching San Carlos a year-and-a-half later, on 11 April 1853.

Unlike Humboldt and Wallace, Spruce did not just pass through San Carlos, tarrying for only a few days. He made San Carlos his headquarters for 20 months, making several long excursions to the north and east, but staying in the village for periods of five-and-a-half, three and three months. During these stays, he made a number of short excursions in its vicinity to collect plants. In a letter to Sir William Hooker of 27 June 1853, Spruce wrote:

"The gratification I naturally feel at finding myself fairly in in terra Humboldtiana is considerably lessened by various untoward circumstances, not the least of which is the very great difficulty experienced here in procuring the necessaries of life, so great indeed that it occupies nearly all a person's time, especially when the river is filling, and we think ourselves well off at San Carlos when we can eat once a day. Anciently when there were missions in most of the pueblos on the Orinoco and Rio Negro, travellers had in them a ready resource; but for some twenty years past there has not been a padre resident in the Canton del Rio Negro, and scarcely one on the Orinoco out of Angostura. A country without priests, lawyers, doctors, police, and soldiers is not quite so happy as Rousseau dreamt it ought to be; and this in which I now am has been in a state of gradual decadence ever since the separation from Spain, at which period (or shortly after) the inhabitants rid themselves of these functionaries in the most unscrupulous manner ..."[23]

The week after writing this, Spruce and two Portuguese families had to barricade themselves in a house in San Carlos to protect themselves during a drunken festival. In the same letter, Spruce wrote of Don Diego Pina:

"Don Diego is perhaps the only white now living in the Canton del Rio Negro who recollects Humboldt in Venezuela. He was making turtle oil on the Orinoco, on a playa near the mouth of the Apuré, when that distinguished traveller passed on his way towards the cataracts. A person died in San Fernando two or three years ago who had seen Humboldt and Bonpland at Esmeralda, and remembered the difficulty they had in procuring the flowers of the Juvia (Bertholletia excelsa), for which, said he, they offered an ounce of gold. At the season of fruit of this tree the Guaharibos descend much below the raudel [torrent] in order to collect it for food, and at that time the Indians of the Casiquiari, in parties of not

more than five or six, lie in wait for them and carry off such as they can lay hold on, making of them slaves for cultivating their conúcos [manioc fields]. Many Indians on the Casiquiari can show lance-wounds received from the Guaharibos in these expeditions."[24]

Five months later, 23 November 1853, he wrote to George Bentham:

"My collections are very poor, and I am leaving a single case to be dispatched to the Barra by the first opportunity. Even had not my time been so much taken up by hunting up lazy and drunken Indians to their work and seeing they kept to it, I could not have done much in the wet season, when scarcely any trees flower, and there are not ferns here as at Saõ Gabriel, to keep me in work. Besides, the river-side vegetation has hardly any plants not already gathered, either on the Rio Negro or the Lower Uaupés. But among few plants I have gathered there are several interesting for their anomalous structure. The other day on the Casiquiari I gathered a tree, allied perhaps to Ochthocosmus (Ternstromiaceae) [Ternstroemiaceae; *Ochthocosmus* is now placed in the *Ixonanthaceae*], *but approaching also Humiraceae, Olacaceae, and Ebenaceae. I have some others which have something in common with the three orders just named, but do not very clearly belong to any one of them: a new genus of Rhizoboleae [= Caryocaraceae] allied to Anthodiscus [Caryocaraceae], but scarcely combinable with it; a fine series of Dimorphandras [Fabaceae] apparently all undescribed; more nutmegs [Myristicaceae] and Commianthi [Commianthus = Retinophyllum]; and several other things which I doubt not will interest you, if they only reach your hands in safety."*[25]

From Peru three years later, Spruce wrote again of San Carlos:

"At San Carlos the dampness exceeded what I had experienced at Saõ Gabriel and on the Uaupés. If I were writing and chanced to drop a piece of paper on the ground, if I did not take it up for five minutes it was so moistened as not to bear writing on. Specimens well dried and put away in a box would be covered with mould in a month's time; but if left on the table, a single night sufficed to mould them. Any article of metal or ivory left all night on the table would be wet in the morning."[26]

On 9 March 1861, Spruce wrote to Bentham from Ecuador:

"My mode of working is this. When I bring home freshly-gathered plants, I make notes on them in books prepared for the purpose, and add numbers. If any plant seems strange to me, I keep flowers, etc., in water to await a spare interval when I can analyse them microscopically. So soon as the plants are dried I pack them into other paper and add the labels from my notes. As it often happens that, at each packing, I have not two plants of even the same natural order, the risk of transposition is very

small. Indeed, so completely does the reading over of my notes recall the features of the plants, that I feel sure if I were shown the whole of my plants classified in your herbarium, and on blank paper, I could, from consulting my notes, put to them the proper numbers and localities without making perhaps a single mistake. As to positive errors of observation, I am as liable as any other mortal. I would wish to speak with all modesty on that head; and working often in boats, or in dismal huts where a squall would suddenly enter the open doorway and disperse both specimens and labels, there must occasionally have been some transposition of both in gathering them up again. This risk of the blowing away or dropping out of labels was, in fact, what made me give up putting labels to the plants as they were drying."[27]

In spite of botanizing under such conditions, Spruce gathered about 875 collections in Venezuela, about 140 of them from San Carlos and its vicinity; Humboldt and Bonpland made approximately 500 collections in Venezuelan Amazonas.[28] Unfortunately we will never know how many were taken by Wallace.

Spruce left San Carlos for the last time on 23 November 1854, paddling south for Barra. He spent another nine-and-a-half years in South America, most of it in Peru and Ecuador. He embarked from Payta, Peru on 1 May 1864 and sailed to England via the isthmus of Panama, arriving in England on the 28th May, broken in health, having been away 10 days short of 15 years.

Epilogue

Upon leaving San Carlos, Humboldt also continued travelling widely in the Americas for another three years and did not return to Europe until August 1804. Wallace, on the other hand, retraced his steps south and east, and soon returned to England. Spruce spent much more time in the field than either Humboldt or Wallace, yet today his name is far less known than theirs, even by bryologists; this may be because of the wealth of popular publications by Humboldt and Wallace and the lack of any by Spruce. Humboldt's *Personal narrative* is a classic of travel literature; it inspired Darwin to visit the tropics, and his own travel narrative[29] was modelled on Humboldt. Besides the *Personal narrative*, Humboldt published numerous other works as a result of his travels, which unfortunately for him used up his considerable fortune. The *Personal narrative* is still well worth reading. On the other hand, I find Wallace's *Narrative of travels* somewhat less interesting than Humboldt's. Indeed, only about 500 of the 750 copies of the first edition were sold in 17 years, the remaining copies being issued with a new title page by a new publisher in 1870.[30] Wallace is better known for his other books, produced over a 60-year period. Of course, it helped that he was regarded as the co-discoverer with Darwin of the principle of natural selection.

Although Spruce published numerous scientific papers,[31] some of them substantial, the book of his travels did not appear until 15 years after his death, edited by his old friend Alfred Wallace. Had it appeared 40 years before, upon his return from South America, it would have immediately established Spruce in the pantheon of popular nineteenth century travel writers. Spruce's book *Notes of a botanist* is much more readable than Wallace's own *Travels* but it was published too late to bring Spruce popular fame.

Alexander von Humboldt, Alfred Russel Wallace, and Richard Spruce helped make known to the rest of the world one of the remotest places in the American tropics. Even today, there is less than one person per square mile in the Venezuelan province of Amazonas. It is still almost as difficult to travel overland there as it was in their times. Most travel is still being done along natural waterways, albeit today by motorised craft, not by the muscles of native canoeists. The legacy of these three great nineteenth-century explorers is their collections and writings, with which our modern understanding of San Carlos and its surroundings begins.

NOTES

(1)   C.F. Jordan (ed.), *An Amazonian rain forest: the structure and function of a nutrient stressed ecosystem and the impact of slash-and-burn agriculture,* Paris, 1989.

(2)   A. von Humboldt, *Personal narrative of travels to the equinoctial regions of America during the years 1799-1804. By Alexander von Humboldt and Aimé Bonpland* (transl. & ed. T. Ross), London, 1852, vol. 1, ix-x.

(3)   Ibid., vol. 1, 136.

(4)   Ibid., vol. 2, 371.

(5)   Ibid., vol. 2, 330.

(6)   Ibid., vol. 2, 390-1. *Bertholletia excelsa* was first described by Humboldt and Bonpland in their work *Plantae aequinoctiales per regnum Mexici ... ad Oronoci, fluvii Nigri, fluminis Amazonum ripas nascentes*, 1808 [1807], vol. 1, 122-7, pl 36.

(7)   Humboldt, op. cit. (2), vol. 2, 356.

(8)   N. Papavero, *Essays on the history of neotropical dipterology, with special reference to collectors (1750-1905).* São Paulo, 1971, vol. 1, 38.

(9)   W.H. Edwards, *A voyage up the river Amazon, including a residence at Pará,* New York, 1847.

(10)   A.R. Wallace, *A narrative of travels on the Amazon and Rio Negro, with an account of the native tribes and observations on the climate, geology and natural history of the Amazon valley*, 2nd edn, London, 1889, ix.

(11)   H.W. Bates, *The naturalist on the river Amazons*, London, 1863, vol. 1, iii.

(12)   Wallace, op. cit. (10), 160-161.

(13)   A.R. Wallace, *Palm trees of the Amazon and their uses*, London, 1853.

(14)   Wallace, op. cit. (10), 271.

(15)   Ibid., 272.

(16) H.L. McKinney, 'Introduction', in *A narrative of travels on the Amazon and Rio Negro* (A.R. Wallace), New York, 1972, xi.

(17) Wallace, op. cit. (10).

(18) Wallace, op. cit. (13).

(19) R. Spruce, *Notes of a botanist on the Amazon and Andes* (ed. A.R. Wallace), London, 1908, 2 vols.

(20) Ibid., vol. 1, xxxii-xxxiii.

(21) Ibid., vol. 1, 2-4.

(22) Ibid., vol. 1, 72, 75.

(23) Ibid., vol. 1, 357.

(24) Ibid., vol. 1, 356.

(25) Ibid., vol. 1, 380.

(26) Ibid., vol. 1, 301.

(27) Ibid., vol. 1, 314-5.

(28) O. Huber and J.J. Wurdack, 'History of botanical exploration in Territorio Federal Amazonas, Venezuela', *Smithsonian Contributions to Botany* (1984), **56**,1-83.

(29) C.R. Darwin, *Journal of researches into the geology and natural history of the various countries visited by HMS Beagle under the command of Captain FitzRoy, R.N., from 1832 to 1836*, London, 1839.

(30) McKinney, op. cit. (16).

(31) Seaward (Ch. 27), this volume, 303-314.

6

Bates, Wallace and economic botany in mid-nineteenth century Amazonia

John Dickenson

Bates, Wallace and economic botany in mid-nineteenth century Amazonia

John Dickenson

University of Liverpool, Liverpool, UK

Introduction

In contrast to their colonialist success elsewhere in the tropics, in South America the British secured only the toehold of Guiana. Nonetheless, they were remarkably active in the continent's politics and economy in the nineteenth century. In the specific case of Brazil, A.K. Manchester could write that *"Until 1914, British capital, British enterprise, British shipping, and British goods predominated in the economic life of Brazil"*.[1] After the opening of the ports, and the country, to foreigners in 1808, British visitors observed and wrote about Brazil's landscape, nature, and society. However, with the exception of the landfall of the 'Beagle' and Charles Darwin, there were no British 'expeditions' to compare with those of the French Castelnau [1843-47], Austro-Germans Pohl, Natterer, Spix and Martius [1818-20] and Adalbert [1842], or the Swiss-Americans Agassiz and Hartt [1865-66]. Instead, we have the commentaries of solitary travellers - diplomats, naval officers, railway and mining engineers, governesses, businessmen, and naturalists. Among the latter we may note Waterton, Burchell, Gardner, Spruce, Bates and Wallace.

The role of such naturalist-travellers is significant for, as the Brazilian ecologist Maria Pires-O'Brien has recently noted, *"the scientific literature of Brazilian natural history before 1900 shows that most practitioners were either foreign travellers or expatriates"*.[2] Their importance extends beyond natural history *per se* for their narratives provide insight not merely into the wildlife of Brazil: they reflect a scientific interest in the flora and fauna of the country, but also a commercial interest, recording the potential and actual value of these resources. In addition, they detail the nature of the agricultural development which had already taken place, using indigenous crops and those imported from the Old World.

From Amazonia, we have the published reports of around a dozen British travellers, who were engaged in a variety of activities. These included the delimitation of boundaries[3] and exploration to acquire basic topographic knowledge of the area. In this latter case, the search for a route via the Andes and the Amazon and its headwaters, to link the Pacific and Atlantic and avoid the long route around Cape Horn, was significant in British exploration of Amazonia. It engaged Maw in 1828 and Smyth and Lowe in 1835 and later, with pressure to open up the river to international navigation in the 1860s, fostered William Chandless's explorations of numerous south bank tributaries.[4] Commercial motives were significant - to identify (and occasionally remove) resources which might be useful to British industry, to create markets for British goods, or to facilitate such commerce by improved transport links. Such activity is exemplified by the travels of Brown and Lidstone, at the behest of the Amazon Steamship

Company.[5] The collection and identification of the region's flora and fauna provide a scientific element in these travels and accounts.

It can be suggested that this scientific activity overlapped with those of exploration and commerce. The scientific travellers recorded not only the plants and animals which they encountered in the forest, but noted their exploitation and use by the native population. They were engaged in economic botany, described as *"the study of plants either useful or harmful to people"*, and in its subdiscipline ethnobotany - *"the investigation of plants employed by people indigenous to a particular area"*.[6] However, it is important to recognise that even 140 years ago Amazonia was not a pristine, virgin forest; it had been modified by Amerindians, Portuguese, African slaves and freemen, and European immigrants. Commentaries thus extended to observations on the collecting and cultivation practices of these settlers, and to what was perceived to be the agricultural potential of these tropical lands. The nineteenth century British travellers saw Amazonia as perhaps unrivalled in its capability to generate a large return for agricultural labour, to yield a great variety of valuable products, and to possess *"all the natural requisites for an immense trade with all the world"*.[7]

Richard Spruce, Alfred Russel Wallace and Henry Walter Bates were the dominant figures in British scientific exploration in Amazonia in the middle of last century. The three naturalists were contemporaries in Amazonia, Bates and Wallace arriving in Belém in May 1848, fourteen months before Spruce. Wallace worked in the region for four years, Spruce for six, before moving to the Andes, and Bates for eleven (Fig. 1). They are a somewhat unlikely trio in the annals of

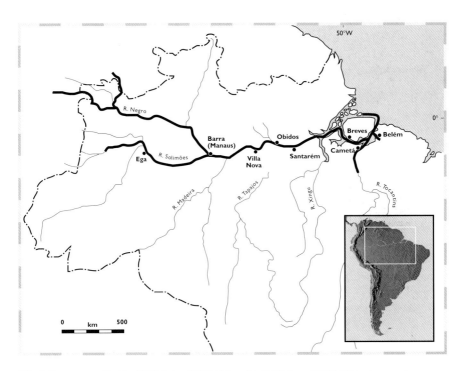

Fig. l. Routes travelled by H.W. Bates (1848-59) and A.R. Wallace (1848-52) in Amazonia.

nineteenth century British explorer-science. They were all of modest origins and education and, as naturalists, were self-taught. They were all 'provincial' - from Leicester, Usk, and Ganthorpe, they were self-financing and essentially independent travellers.

Spruce's *bona-fides* as a botanist is clearly established, from his work in Yorkshire, the Pyrenees, and in Amazonia and the Andes. His contributions to economic botany are also evident, from his role in the collection and removal of *Cinchona* from Ecuador, and his early description of rubber.[8] Moreover, in his report on *Cinchona*, Spruce observed *"I have seen enough of collecting of the products of the forest to convince me that whatever vegetable substance is needful to man, he must ultimately cultivate the plant producing it"*.[9]

Bates and Wallace are more commonly thought of as zoologists, from their insect collecting; from Bates' work on mimicry; and from Wallace's contribution to ideas on evolution. Bates was an entomologist from his earliest days as a naturalist, publishing a precocious paper on the Coleoptera of Bradgate Park at the age of 18.[10] Wallace, in his autobiography, implies that it was flowers, shrubs and trees which first aroused his curiosity in nature, but that he was unaware of *'such a science as systematic botany'*.[11] Though by 1843 he was attempting to classify the plants he observed, he noted *"the time I gave to the study before I left England was not sufficient for me to acquaint myself with more than a moderate proportion of the names of the species I collected"*.[12] Under Bates' influence he became more interested in insects, and the declared objectives of the pair, when they decided to abandon their humdrum jobs as surveyor in Neath and brewer's clerk in Burton-on-Trent respectively for the mysteries of the Amazon rain forest, were *"to make for ourselves a collection of objects, dispose of the duplicates in London to pay expenses, and gather facts ... towards solving the problem of the origin of the species"*.[13]

Yet, despite their seeming commitment to zoology, they found time in the brief period before their departure for Brazil in the spring of 1848, to inquire of Sir William Hooker as to which plants Kew would like them to collect. It might be significant that 1848 was also the year in which Kew opened its Museum of Economic Botany.[14] They also visited Chatsworth to inspect types of orchids they might encounter in the rain forest.[15] However, this visit to the orchid houses appears to have been of limited utility, for Bates observed that *"orchids are very rare in the dense forests of the lowlands"*[16] and Wallace that *"here there are none but a few small species with dull brown or yellow flowers"*.[17]

The bulk of their scholarly writings, with the exception of *Palm trees of the Amazon and their uses*,[18] are concerned with fauna rather than flora, but both write enthusiastically of their first sight of *"the great primaeval forest"*, on May 26th, 1848.[19] Wallace describes the dense forest, palms and plantains, which appeared "doubly beautiful from the presence of those luxuriant tropical products in a state of nature, which we had so often admired in the conservatories of Kew and Chatsworth".[20]

Collecting in economic botany

Despite the commitment to entomology, there is evidence in Bates' published correspondence that he did engage in the commercial collection of botanic material. In his letters to his agent Stevens and others, published in *The Zoologist*, there is reference to economic botany. However, of the 40 letters which span Bates' period in Amazonia, only seven refer, generally fleetingly, to this activity. In a letter of 1852 Bates wrote: *"I hope some of my specimens of medicinal and economic botany may prove curious and saleable"*.[21] He refers to preparing *"a few chests of botanical curiosities"* at Belém in April 1851[22] and to the despatch of specimens of economic and medical botany from Santarém in November 1852 and April 1854.[23] However, three months labour at Villa Nova in late 1854 yielded only two items, *"not worth sending by themselves"*.[24]

There is little detail as to the items sent. The Belém shipment consisted of *'young palms dried, curious fruit &c.'*,[25] while the modest items from Villa Nova were Piao and the silky fibres of a species of Malvaceae. The latter at least indicates the object of the exercise, for Bates suggests that "it might be made a branch of commerce".[26]

There are clues that Bates was collecting specifically, sending material via Stevens to Messrs Saunders and Hanbury. The latter is thanked for *"The estimation bestowed on my miserable collection of Economic Botany"*, and for pamphlets and hints received.[27] Bates claimed that he never lost *"an opportunity of acquiring objects in* [Hanbury's] *department"*.[28] Balsams, resins, and medical barks and roots appear to have been of particular concern, though Bates implies that some of the balsams requested might be difficult to obtain - *"I scarcely expect to find any of the different kinds of balsams of Peru in the valley of the Amazons, except it is near the eastern foot of the Cordillera"*.[29]

It is also evident that as a novice botanist, Bates did not find this work easy. In 1852 he noted that a collection of woods was awaiting the flowers,[30] and later observed that while it was possible to collect various kinds of balsams, resins and medicinal plants, *"the real difficulty is in identifying those separate objects with the tree which produces them, and acquiring a flowering specimen of it. This is much aggravated by the loose terminology of the Indians, who give the same name to very different things"*.[31]

More broadly, he recorded several obstacles to collecting in economic botany - that the lack of a timber trade made it hard to obtain wood samples; that even with an illustration, it was difficult to find a particular species in the vast variety of the forest; and that few people could recognise particular trees. Even if the desired species was found, it might not be in flower or fruit, and probably the guide would be unable to indicate when this might occur. Significantly perhaps, Bates observed that *"I despair of getting up the illustrative series of Economic Botany complete"*.[32]

The lesser number of Wallace's letters published by Stevens in the *Annals and Magazine of Natural History* and *The Zoologist* contain minimal reference to botany. His observations on economic botany are encompassed in the postscript chapter to *A narrative of travels on the Amazon and Rio Negro*, on the vegetation of Amazonia.[33] This includes discussion of 'useful plants' - hardwoods, india-rubber, Brazil-nuts, sarsaparilla, nutmeg, tonquin beans, sassafras oil and Cravo de

Maranhão bark. Including the above, he provides a list of seventeen 'principal vegetable productions of commercial value' in the forest.[34]

Beyond these specific discussions of collecting in economic botany and reference to useful vegetation, we can also glean information on the potential and utilisation of Amazonia's vegetation from the travel narratives of Bates and Wallace. Both make considerable comments not only on the plants they observed but on their gathering and cultivation, and their use for subsistence or commerce, by Indians, mamelucos, negroes or Europeans. In fact many of these vegetable products had multiple uses - for native and European subsistence, and for commerce. In broad terms it is possible to examine the vegetable products recorded, their use, and their location.

Gathered plants

The various plants mentioned by Bates and Wallace as being gathered from the forest, and the uses to which they were put, fall into a number of categories.

A. Foodstuffs. These were primarily fruits. At Ega, Bates commented that *"Fruits of the ordinary tropical sorts could generally be had. I was quite surprised at the variety of the wild kinds, and of the delicious flavour of some of them. Many of these are utterly unknown in the regions nearer the Atlantic".*[35] He provides comments on the taste of these 'particular productions', which included Jabuti-púhe, Cumá , Pamá, Umarí, Wishí, Purumá, Uiki, Wajurú, Cashipári-arapaá, two kinds of Bacuri, and the peach palm.[36] Elsewhere he notes Aápiránga, Abiús, Alligator pear, Atta, Capú-ai, Genipapa, Guava, Pikiá, and Urucurí. To this list Wallace adds Berribee, Marajá, Miriti, Ocokí, and Puxuri. There are also several references to Sapucaya, Brazil, and cashew nuts. The diverse utility of these forest products is indicated by Bates' description of the cashew as source of fruit, roasted nut, and 'wine' to cure skin disease, and of the Umirí as yielding fruit not unlike the Damascene plum in taste and *"an oil of the most recherché fragrance"*, highly esteemed by the native women.[37]

The continuing role of Amazonia as a source of spices, first exploited by the missions, is evident in Wallace's listing of nutmeg, Cravo [clove], Canella [cinnamon], and Vanilla.[38] Indian use of a diversity of plants to make beverages is also indicated. Sources included Assaí, Baccába, Patawa palm[39] and guaraná. Bates identified the Munducurú as the only tribe who manufactured guaraná, selling it in large quantities to traders as a remedy for diarrhoea and fever.[40]

The collection of sarsaparilla is reported by both men, in various locales - along the Tocatins, Tapajós and Negro, and traded through Barra, Santarém, and Belém.[41] Wallace provides a description of the plant and its packaging for transport.[42] The role of the Indians as collectors was significant - Bates recorded that the Munducurús on the Tapajós *"gather large quantities of salsaparilla".*[43] Its significance in the gathering economy is evident in his comment (on the Solimões near Ega) that its production was *"long ago exhausted in the neighbourhood of towns, at least near the banks of the rivers ... and* [it is] *now got only by more adventurous traders during long voyages up the branch streams".*[44] He also notes its use to treat skin disease among the Indians.[45]

B. Timber. Their earliest forays outside Belém were to the timber mill of Mr Leavens at Magoary and on his search for cedar on the Tocantins. The mills used Pao d'Arco and Massaranduba. Both note the utility of Massaranduba, not only as a source of sawn wood, but of juicy fruit, durable glue, and milk for tea, coffee and custard.[46] Upstream, the forest of Altar do Chão yielded choice woods - laurel and Itauba for shipbuilding, tortoise-shell wood for walking-sticks, Moira coatiára for cabinets, Sápu-píra for mortars, along with ebony and marupá.[47]

In a classic piece of ethnobotany Wallace describes their negro cook's identification of Caripé for pottery making, Ocoóba for medicine, Pootiéka for paddles, Nowará for charcoal, and Quaroóba for house timber.[48]

C. Fibres. These included the silk-cotton tree [*Magabeira*] and especially the Piassába, which was an important trade item at Barra. Wallace described its appearance, site, collection and use, noting that *"a great part of the population of the Upper Rio Negro is employed in obtaining the fibre for exportation".*[49] He gives precise details of its distribution, on the banks of several rivers, though not on the banks of the Negro itself, but on only three of five north bank tributaries, and two of five on the south.

D. Other plants. Unsurprisingly, Wallace comments on a variety of palms and their uses, including Assai for palm heart, fruit and drink, Inajá as a vegetable, and urucurí, used in the processing of rubber.[50] Similarly, Bates notes the fruits of the Assai, Mirití, and Macujá, and use of the leaf stalks of the Jupatí to make sails.[51] As a footnote to this theme, Bates, who complained frequently of his lack of reading material, particularly scientific, noted in 1854 that he now had two copies of Wallace's *Palms "and that it is really a very correct, useful book on the class. I can add many species, however, to his list, and I doubt not Mr Spruce could double it".*[52]

Forest sources of medicinal plants include sassafras, Súcu-úba, and Mururé and, for insecticides, andiroba oil and Breio branco. Related forest products included the Tonka bean, for scenting snuff, of which between 1000 and 3000 lbs were exported via Santarém from the Tapajós; the use of Paricá snuff as a stimulant and for medicinal purposes by the Múras and Mauhés; and, on the Solimões, the use of Ypadú (coca).[53] Both Bates and Wallace describe the use of timbo root in fishing.[54] Collection of balsam of copaiba was evidently widespread along the Madeira, Tapajós, and Solimões. The use of various vegetable dyestuffs, for fibres, pottery, and body painting was also noted, with Bates observing use by the Tupi of Comateu, Urucú and indigo to paint gourd drinking cups.[55]

Such commentaries are in part an indication of the natural potential of the forest, but also derive from observations of Indian use of its resources. There are numerous specific references to the latter, particularly in Wallace's appendix on the aborigines, which details crops, weapons, dwellings and utensils.[56] For the tribes of the Uaupés, he lists the cultivation of manioc, sugar, sweet potato, yams, maize, pineapples, plantains and bananas, peppers, cashews, tobacco, cocura, pupunha palm, urucú, abios, ingás, plants for dyes and cordage, and palms for beverages.[57] In addition, he gives the specific sources of wood and fibre used for bows and arrows, and the forest products used for dwellings, furniture, canoes, weapons and clothing. It is also worth noting that Wallace records Indians trading sarsaparilla, pitch, string, baskets and hammocks for European goods, of which sarsaparilla was *"by far the most valuable product, and is the only one exported".*[58] Interestingly, even 150 years ago Wallace recorded the inter-tribal exchange of European goods for sarsaparilla with Indians of remote districts, such that *"numerous tribes, among whom no civilised man has ever yet penetrated, are well supplied with iron goods, and sent the product of their labour to European markets".*[59] Bates provides similar, though less detailed, commentary on a number of Indian groups, such as the Munducurú and Túcana.[60]

E. Rubber. Amazonia's most famous forest product deserves particular mention. Its place in economic botany is well known, at least as far as the tale was polished by Henry Wickham.[61] This frequently repeated version has recently been called into question by Warren Dean, suggesting that Wickham was perhaps elastic with the truth.[62] Certainly, well before 1876, Spruce, Bates and Wallace had all provided comment on the india-rubber economy. Dean notes that Spruce was the first to describe the techniques of rubber tapping and processing,[63] but both Bates and Wallace make reference to the distribution and processing of rubber (Fig. 2).

Wallace notes its collection at various points along the Tocantins in 1848, and at Breves,[64] and records their excitement at cutting a seringa and seeing its juice flow, at Magoary in July, 1848.[65] Later, on the Tocantins, he describes its daily collection from small incisions into the bark, its moulding into flat cakes, shoes, and bottles, and its smoke-hardening.[66] In his more general review of the region's vegetation, he indicates that the main area from which it was collected

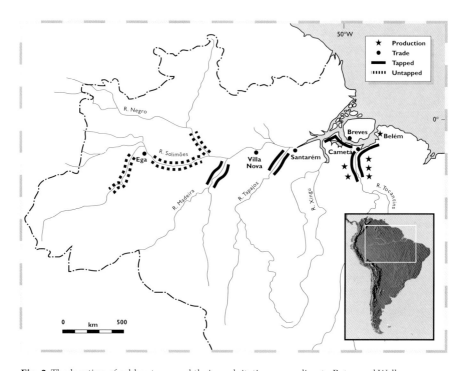

Fig. 2. The location of rubber trees and their exploitation, according to Bates and Wallace.

was between Belém and the Xingú, although it was also found, but not yet collected, on the upper Amazon and the Negro.[67]

Bates' comments are fewer, but geographically more dispersed, noting its trade at Cametá, Vila Nova, Santarém and Ega, and its collection on the Tocantins, Madeira, Tapajós and Solimões. By the mid-1850s he records india-rubber as the chief export of Belém.[68] The beginning of the rubber boom is evident in his comment that while traders visited the Madeira up to 1853 to collect rubber and other products, *"in that year many indiarubber collectors resorted to this region, stimulated by the high price (2s.6d. a pound) which the article was at that time fetching at Pará"*.[69] Of his later residence at Ega he noted that the *"search for India-rubber has commenced but very lately"*.[70]

Bates provides detailed descriptions of rubber tapping and processing, and of its collection on the islands and swampy lowlands 50 to 100 miles west of Belém. However, *"there are plenty of untapped trees still growing in the wilds of the Tapajós, Madeira, Jurúra and Jauarí, as far as 1800 miles from the Atlantic coast"*.[71] It is interesting that some of these observations pre-date Spruce's arrival in Amazonia, but were not published until 1853 and 1863 respectively (as against Spruce's 1855 paper). In his account of the Solimões, published in 1852, Bates described the Juruá as *"teeming with valuable productions"*, including india-rubber.[72] Spruce and Bates certainly spent more time and explored the region more thoroughly than did Wickham on his cruise down the Negro and Amazon in 1870 and his residence near Santarém in the early 1870s.

After his return to England in 1864, Bates became Assistant Secretary of the Royal Geographical Society, described by one of his biographers as 'an unofficial extension of the Foreign and Colonial Offices'.[73] One of the explorers he dealt with was William Chandlass, whose reports on the southern tributaries of the Amazon in the 1860s, and correspondence at the RGS, contain reference to the distribution of rubber trees - all before Wickham's letter to Hooker of 1872, or that of Clements Markham to Hooker of 1873.[74] Markham had close links with Spruce in the removal of *Cinchona* from the Andes in 1860-62; in 1864 Markham was involved in Bates' appointment to the RGS and was his superior at the Society, as its Honorary Secretary or President, until Bates' death in 1892. Though no written evidence has yet come to light, it seems improbable that the two men did not, between 1864 and 1876, discuss the rubber resources of Amazonia and their distribution.

New World cash crops

The greatest significance in economic botany is, of course, the economic use of local crops. By the middle of the century Bates and Wallace had demonstrated a considerable use of New World plants in subsistence and commercial agriculture, by natives, caboclos, negroes, and Europeans. Such activity covered a range of crops, and was evident along the extent of their routes.

The most significant cash crop appears to have been cacao, recorded along the valley from Belém to Ega, in small groves and large plantations, for local use and for export. It was described as 'abundant' in some places, with substantial plantations being cultivated - of 8,000 trees at Serpa and 60,000 at Jambouassu.[75]

Bates commented, however, that though cacao was the staple crop of the Óbidos district, the plantations were old and ill-cared for, with poor processing, thus off-setting its good natural quality, and diminishing its price.[76] For a low yielding cacaoal of 10,000 trees at Cametá Bates quotes a return of £40.[77] Cacao was evidently not only a crop significant in regional land use, but in trade along the river, and in the commercial economies of Cametá and Ega.[78]

Cotton and tobacco receive only passing reference and seem to have been cultivated only on a small scale, though the latter was evidently a trade crop.[79] Wallace provides a detailed outline of the planting and cultivation of the crop, and of its harvesting, drying and manufacture into cigars or, more usually, into 2 to 4 pound rolls.[80] Other indigenous crops included maize, pumpkin, capsicum, chilli, sweet potato, pawpaw and pineapple.

The most important element in the Amazon farm economy was manioc, to which both Bates and Wallace make frequent reference, as to its cultivation by Indians, mamelucos, and Europeans, its distribution, and its processing for farin-ha, mingau, tapioca, tucupi sauce, and fermented beverages. Wallace gives a detailed description of its processing, and describes farinha de mandioca as glutinous, looking like sawdust, but very nutritious.[81] Scale of production obviously varied enormously - from plantations using 150 slaves to small patches cultivated by women - with very large plantations on the Tapajós yielding an annual surplus for sale by the Munducurús of 8 to 14 tons.[82]

Old World imports

The focus of economic botany is on the potential use of the vegetable produce of a particular area, and an important element in the work of Spruce, and to a lesser extent Bates and Wallace, was to identify and comment on such resources in Amazonia. These formed part of the botanical impact of the New World on the Old. However, as Viola has recently noted, the 'Discoveries' in 1492 and 1500 saw 'seeds of change' moving in two directions, between Old World and New.[83] In the writings of Bates and Wallace we can find numerous references to Old World crops evidently introduced, and cultivated, in Amazonia. These include coffee, sugar, and rice as subsistence and cash crops, and a range of fruit and vegetables, including oranges and other citrus fruits, bananas and plantains, yams, mangoes, tamarind, water melon and even 'cabbages and onions introduced from Europe'.[84]

The British, in their efforts to settle around the mouth of the Amazon, had planned to 'plant sugar canes and to erekt ingenies for makinge suger' before 1630.[85] Two centuries later Bates and Wallace recorded sugar cultivation at locations from Belém to Ega, sometimes as part of Indian agriculture,[86] or as substantial plantations with mills and distilleries.[87] Wallace records one plantation, on the Guamá river, worked by slave labour, as being 1.5 by 0.25 miles in extent,[88] while Bates provides details of the rich alluvial soil, the length of the cane, and the mill - 'a rude affair, worked by bullocks' of another, at Óbidos.[89] He also refers to sugar-milling on the Mojú, where a local rebellion had reduced the number of mills from eleven to three.[90] Frequent reference is also made to the distillation of cane juice into cachaça.

Coffee was a later, eighteenth century, introduction from Cayenne,[91] but by the 1850s it seems to have become widely cultivated along the Amazon - indeed Wallace observed near Belém 'on almost every roadside, thicket, or waste, the coffee-tree is seen growing'.[92] Both make frequent reference to its cultivation, usually in groves, or as part of the farm economy, but occasionally in plantations.[93]

Rice was a common staple, but there are a few references to larger scale production and processing. Wallace describes the American-managed rice-mills at Magoary, one powered by water, the other by steam, processing rice grown by Indians and small land-holders.[94] Another mill, on the Capim, he describes as 'one of the best modern buildings I had seen in the country', built at a cost of several thousand pounds.[95]

Bananas form part of the rural diet, for Indians, mamelucos, and Portuguese, and cultivation of oranges and other citrus fruits appeared to be widespread. However, the impact of the rubber boom is evident in Bates' comment on the eve of his departure from Belém, that *"Oranges, which could formerly be had almost gratis, were now sold in the streets at the rate of 3 for a penny"*.[96]

The role of immigrants in introducing Old World agricultural crops is exemplified by a Portuguese settler on the Solimões described by Wallace as having planted *"oranges, tamarind, mango and many other fruit-bearing trees, made pleasant avenues, gardens, and pastures, stocked them well with cattle, sheep, pigs and poultry, and set himself to the full enjoyment of a country life"*.[97]

What is evident from these observations is the extent and diversity of the Amazon farm economy 150 years ago, with a variety of crops, local and imported, cash and subsistence. Thus Wallace describes a farm near Juhatí cultivating cotton, tobacco, cacao, manioc, and bananas, and raising pigs, ducks and chickens,[98] whilst Bates records an Indian farm at Barreiros Cararaucú as having *"a large plantation of tobacco, besides the usual patches of Indian-corn, sugar-cane, and mandioca; and a grove of cotton, cacao, coffee and fruit trees"*.[99] On the Tapajós the Munducurú not only gathered sarsaparilla, rubber, guaraná, and Tonka beans and cultivated manioc and tobacco, but had incorporated sugar into their economy.[100] Whether gathered from the forest or cultivated as New or Old World crops, a complex agricultural economy in Amazonia had emerged by the mid-nineteenth century. By 1860 there was already a heterogeneity of products introduced from Europe intermingled with the 'wonderful variety of the tropics', although, as Bates commented, *"It must not be supposed that these plantations and gardens were neatly kept. Such is never the case in this country, where labour is scarce"*.[101]

Economic potential

Bates and Wallace wrote, in a distinctly Victorian manner, of the unrealised potential of Amazonia, where the virgin land contained *"timber-trees in inexhaustible quantities, and of such countless varieties that there seems no purpose for which wood is required, but that one of a fitting quality may be found"*.[102] In such lands 'sugar, cotton, coffee and rice might be grown in any quantity and of the finest quality,[103] Wallace saw the region as an 'earthly paradise', where *"the "primeval" forest can be converted into rich pasture and meadow land, into cultivated fields, gardens, and orchards, containing every variety of produce"*.[104] In his view, the Amazon could sustain abundant, luxuriant, and high quality production of both Old and New World crops - coffee and cacao, manioc and rice, cashews and sugar, avocados and oranges.[105] For this to be realised, however, what the region of course required was *"a few families of agricultural settlers from Northern Europe"*.[106]

Conclusion

Neither man returned to the Amazon and their later careers were directed elsewhere - Wallace to Malaysia, to evolution, and to his myriad interests in spiritualism, land nationalisation, and vaccination. Bates' later scientific work was exclusively in entomology,[107] though there are hints of a lingering interest in matters botanical. In his editing of the English version of von Hellwald's *Central America, West Indies and South America* he noted that 'the chief alterations have been in reference to Natural History and the Geographical Relations of Faunas and Floras.[108] More significantly, in his contribution to the Royal Geographical Society's *Handbook for travellers*, the Bible of every self-respecting Victorian explorer, he suggested *"In botany, a traveller, if obliged to restrict his collecting, might confine himself to those plants which are remarkable for their economical uses"*.[109]

The references made by Bates and Wallace to economically useful plants in Amazonia were scarcely even the tip of what Balick describes as *"a huge beneficial 'iceberg' of possibilities present in the Amazon flora"*.[110] Whereas the contributions of Richard Spruce in this field, in the exploitation of *Cinchona* and rubber, are well known, those of Bates and Wallace are peripheral and ephemeral. The available published record is far from perfect - constrained by their modest knowledge of botany, the curious form of both *The travels* and *A narrative*, and because their journeys were riverine and their expeditions largely riparian. They offer, nonetheless, a mid-century record of floral potential and diverse agricultural activity along the banks of the Amazon, Solimões, Rio Negro and their tributaries. Given current concerns for the destruction of the rain forest, it is important to recognise that even when Spruce, Bates, and Wallace 'explored' Amazonia 140 years ago, it was already in process of modification by man and his crops.

[Throughout the text botanical names as used by Bates and Wallace have been followed, though they are not consistent in their spelling, or use of capital letters and accents, even regarding, for example, india rubber and sarsaparilla.]

NOTES

(1)   A.K. Manchester, *British pre-eminence in Brazil*, New York, 1972, ix.

(2)   M.J. Pires-O'Brien, 'An essay on the history of natural history in Brazil, 1500-1900', *Archives of Natural History* (1993), **20**, 37.

(3)   R.H. Schomburgk, 'Journey from Fort San Joaquim, on the Rio Branco, to Roraima, and thence by the rivers Parima and Merewari to Esmeralda on the Orinoco, in 1838-9', *Journal of the Royal Geographical Society* (1841), **10**, 191-247.

(4)   H.L. Maw, *Journal of a passage from the Pacific to the Atlantic*, London, 1829; W. Smyth and F. Lowe, *Narrative of a journey from Lima to Para, across the Andes and down the Amazon*, London, 1836; W. Chandless, 'Notes on the rivers Arinos, Juruena and Tapajos', *Journal of the Royal Geographical Society* (1862), **32**, 268-280; W. Chandless, 'Ascent of the River Purûs', *Journal of the Royal Geographical Society* (1866), **36**, 86-118; W. Chandless, 'Notes on the River Aquiry, the principal affluent of the River Purûs', *Journal of the Royal Geographical Society* (1866), **36**, 119-128; W. Chandless, 'Notes on a journey up the River Juruá, *Journal of the Royal Geographical Society* (1869), **39**, 296-311.

(5)   C.B. Brown and W. Lidstone, *Fifteen thousand miles on the Amazon and its tributaries*, London, 1878.

(6)   M.J. Balick, 'Useful plants of Amazonia: a resource of global importance', in *Key environments: Amazonia* (eds. G.T. Prance and T.E. Lovejoy), Oxford, 1985, 360.

(7)   A.R. Wallace, *A narrative of travels on the Amazon and Rio Negro*, New York, 1972, 261.

(8)   L.H. Brockway, *Science and colonial expansion: the role of the British Royal Botanic Gardens*, New York, 1979, 113-114; R. Spruce, 'Note on the India-rubber of the Amazon', *Hooker's Journal of Botany* (1855), **7**, 193-196.

(9)   R. Spruce, *Report on the expedition to procure seeds and plants of the Cinchona succirubra or Red Bark tree*, London, 1861, 83.

(10) H.W. Bates, 'Note on the coleopterous insects frequenting damp places', *Zoologist* (1843), **1**, 114-115.

(11) A.R. Wallace, *My life, a record of events and opinions*, 2nd edn., London, 1908, 61.

(12) Ibid., 104.

(13) H.W. Bates, *The naturalist on the river Amazons*, London, 1892, vii. [Reprint of the unabridged 1863 edition.]

(14) Brockway, op. cit. (8), 83.

(15) H.P. Moon, *Henry Walter Bates, FRS, 1825-1892, explorer, scientist and Darwinian*, Leicester, 1976, 20-21.

(16) Bates, op. cit. (13), 34.

(17) Wallace, op. cit (7), 7.

(18) A.R. Wallace, *Palm trees of the Amazon and their uses*, London, 1853.

(19) Bates, op. cit. (13), 1.

(20) Wallace, op. cit. (7), 1.

(21) H.W. Bates, 'Extracts from the correspondence of Mr. H.W. Bates, now forming entomological collections in South America', *Zoologist* (1849-58), 3899.

(22)   Ibid., 3231.

(23)   Ibid., 3897, 4552.

(24)   Ibid., 5014.

(25)   Ibid., 3231.

(26)   Ibid., 5015.

(27)   Ibid., 4202, 4552.

(28)   Ibid., 5015.

(29)   Ibid., 4552.

(30)   Ibid., 3728.

(31)   Ibid., 4550.

(32)   Ibid., 3899.

(33)   Wallace, op. cit. (7), 300-309.

(34)   Ibid., 305.

(35)   Bates, op. cit. (13), 285.

(36)   Ibid., 285-286.

(37)   Ibid., 190, 253, 191.

(38)   Wallace, op.cit. (7), 304-305.

(39)   Ibid., 338.

(40)   Bates, op. cit. (13), 245.

(41)   Ibid., 66, 184; Wallace, op. cit. (7), 195, 267.

(42)   Wallace, op. cit. (7).

(43)   Bates, op. cit. (13), 243.

(44)   Ibid., 282.

(45)   Ibid., 371.

(46)   Wallace, op. cit. (7), 20.

(47)   Bates, op. cit. (13), 218-219.

(48)   Wallace, op. cit. (7), 22-23.

(49)   Ibid., 167.

(50)   Ibid., 53.

(51)   Bates, op. cit. (13), 62-63, 79.

(52)   Bates, op. cit. (21), 4551.

(53)   Bates, op. cit. (13), 203, 169, 283.

(54)   Wallace, op. cit. (7), 169.

(55)   Bates, op. cit. (13), 114.

(56)   Wallace, op. cit. (7), 331-361.

(57)   Ibid., 336-337.

(58)   Ibid., 349

(59)   Ibid.

(60)   Bates, op. cit. (13), 240-243, 381.

(61)   H.A. Wickham, *On the plantation, cultivation, and curing of Parà Indian rubber (Hevea brasiliensis), with an account of its introduction from the west to the eastern tropics*, London, 1908.

(62)   W. Dean, *Brazil and the struggle for rubber*, New York, 1987.

(63)   Ibid., 10.

(64)   Wallace, op. cit. (7), 38-56, 93.

(65)   Ibid., 24.

(66)   Ibid., 53-54

(67)   Ibid., 303.

(68)   Bates, op. cit. (13), 72-73, 140.

(69)   Ibid., 161.

(70)   Ibid., 282.

(71)   Ibid., 72.

(72)   H.W. Bates, 'Some account of the country of the River Solimoens, or Upper Amazon', *Zoologist* (1852), 3597.

(73)   G. Woodcock, *Henry Walter Bates, naturalist of the Amazons*, London, 1969, 257.

(74)   Chandless, op. cit. (4); Dean, op. cit. (62), 13.

(75)   Bates, op. cit. (13), 163; Wallace, op. cit. (7), 53.

(76)   Bates, op. cit. (13), 121, 139-140.

(77)   Ibid., 79.

(78)   Ibid., 84, 282.

(79)   Wallace, op. cit. (7), 144.

(80)   Ibid., 126-170.

(81)   Ibid., 336-338, 12.

(82)   Ibid., 36; Bates, op. cit. (13), 146, 243.

(83)   H.J. Viola, 'Seeds of change', in *Seeds of change: a quincentennial commemoration* (eds. H.J. Viola and C. Margolis), Washington, 1991, 11-16.

(84)   Bates, op. cit. (13), 163.

(85)   J. Lorimer (ed.), *English and Irish settlement on the River Amazon, 1550-1646*, London, 1989, 62.

(86)   Bates, op. cit. (13), 98.

(87)   Wallace, op. cit. (7), 39.

(88)   Ibid., 86.

(89)   Bates, op. cit. (13), 124.

(90)   Ibid., 84.

(91)   A.C. Ferreira Reis, 'Economic history of the Brazilian Amazon', in *Man in the Amazon* (ed. C. Wagley), Gainesville, Fla, 1974, 33.

(92) Wallace, op. cit (7), 8.

(93) Ibid., 42, 50.

(94) Ibid., 19, 21.

(95) Ibid., 80-81.

(96) Bates, op. cit. (13), 387.

(97) Wallace, op. cit., (7), 129.

(98) Ibid., 45.

(99) Bates, op. cit. (13), 156.

(100) Ibid., 242-243.

(101) Ibid., 163.

(102) Wallace, op. cit. (7), 32.

(103) Ibid., 55.

(104) Ibid., 230-231.

(105) Ibid.

(106) Bates, op. cit. (13), 198.

(107) J.P. Dickenson, 'H.W. Bates - the naturalist of the River Amazons', *Archives of Natural History* (1992), **19**, 209-218.

(108) H.W. Bates, *Central America, West Indies and South America*, 2nd edn., London, 1882, v.

(109) H.W. Bates, 'Hints on the collection of objects of natural history', *Proceedings of the Royal Geographical Society* (1871), **16**, 76.

(110) Balick, op. cit. (6), 360.

7

Richard Spruce and Margaret Mee: explorers on the Rio Negro, a century apart

Ruth Stiff

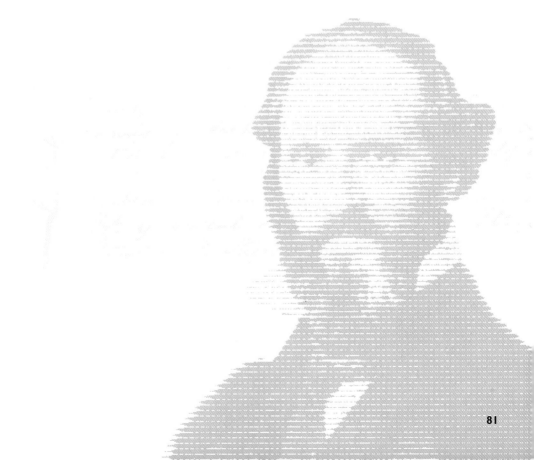

Richard Spruce and Margaret Mee: explorers on the Rio Negro, a century apart

Ruth Stiff

North American Exhibitions Curator,
Royal Botanic Gardens, Kew, UK

As Curator of North American Exhibitions for the Royal Botanic Gardens, Kew (RBG, Kew) I was asked two years ago by the Margaret Mee Amazon Trust (MMAT) to mount an exhibition of Margaret Mee's paintings, drawn from the Collections of the RBG, Kew, that would travel throughout the United States and Canada. Organised by RBG Kew, MMAT and the Houston Museum of Natural Science, this exhibition is expected to open early in 1997.

Fig. 1. Margaret Mee (aged 78, Rio Negro, Amazonia, 1988).

In my research for this exhibition I came across the name Richard Spruce in Margaret Mee's diaries time and time again, as her itineraries often followed those of Spruce, particularly in the Rio Negro region of Amazonia. Professor Richard Schultes, who is on the Board of the MMAT, talked frequently about Spruce when we first met at Harvard University to discuss the Margaret Mee Exhibition. He has been intrigued with the common pursuits of these two intrepid travellers for many years and encouraged me to investigate their common experiences.

Margaret Mee followed in the grand tradition of British exploration. She has been referred to as the premier female explorer of the Brazilian rain forest, an outstanding botanical artist, and acclaimed world-wide by botanists and art critics alike.

She realised her true vocation as an explorer and botanical artist after moving to Brazil from Britain in the 1950s. Initially, she was attracted to the southern coastal mountains (the Serra do Mar) by the lush tropical flora of the region, where she began to paint the native bromeliads of this coastal forest. During this period she worked alongside Dr Lyman Smith, an American botanist from the Smithsonian Institution, who was studying bromeliads at the Instituto de Botânica de São Paulo. With Dr Smith, Margaret produced numerous paintings for a book on bromeliads.[1]

Four years later, however, on her first trip to the great Amazon Basin, she became captivated by its beauty and diversity. Over the next three decades, starting at the age of 47, Margaret made fifteen long and arduous journeys to Brazilian Amazonia to search for, collect and paint the plants of the Amazon forests.

She often travelled with a single Indian guide, learning the ways of the forest while living amongst its plants, animals and people. She would often live for weeks with the Tucano Indians, who were a great source of information about the names of the villages, trees and plants along the river.

Margaret's sketchbook was always at her side. Her working method started with detailed sketches of the plant in the forest, accompanied by diary entries which recorded the plant's colour and habitat, as well as its locality. Carrying her tubes of paint in a small wooden box, she mixed them carefully to achieve exactly the correct intensity of colour.

With the help of native guides, Margaret collected a wealth of fascinating plants, many of which were entirely new to science. She transported her plants in large plastic bags back to her boat, where they were then placed in airy baskets with the roots carefully wrapped in wet newspaper. The plants were brought back to her studio in Rio, where they were used as reference material while the paintings were being worked up into finished compositions. For this reason, Margaret chose to collect plants such as orchids and bromeliads, since they flower readily in cultivation and are easier to collect and transport. These plants were subsequently added to the living collections of various botanical institutions or private estates in Brazil.

Most of Margaret Mee's earlier work is in the classic style, with the image portrayed against a white background, but later, driven by a consuming passion to convey a conservation message through her drawings, Margaret began to use the rain forest environment as a backdrop for her plant portraits. She realised that the significance of her drawings could only be fully appreciated when they were viewed with a dual understanding of both the ecosystems which supported her beloved flowers, as well as the international dilemma caused by the destruction of the rain forest and its prolific diversity of life.

She spoke movingly as a rain forest environmentalist, and her valiant efforts to sensitise world opinion to the endangered state of that rapidly disappearing resource gained her an international reputation. She remained an important spokeswoman for conservation until her death, ironically enough in a car accident in Leicester, England, in November of 1988. An exhibition of her work had just opened at Kew and just six days earlier she had appeared on television on the MacNeil/Lehrer Newshour in the United States; her book *In search of flowers* had just been published.[2]

However, Margaret Mee's legacy lives on. Just a few months prior to her death, the Margaret Mee Amazon Trust was formed. It operates as an active organisation, dedicated to furthering education and research in Amazonian plant life and conservation. In partnership with its Brazilian sister organisation, the Fundação Botanica Margaret Mee, the Trust provides travelling scholarships for young Brazilian artists and scientists, to enable them to study in the U.K. and then return to Brazil to continue Margaret's work. More than twenty such scholarships have been awarded so far.[3]

Margaret Mee's work on Brazil consists of over 400 folio compositions, approximately 40 sketchbooks filled with detailed drawings executed in the field, and 15 diaries. According to Simon Mayo, a botanist with RBG, Kew and an expert on Amazonian flora, these works *"comprise an important scientific archive of the Amazon forests and their flora"*.[4]

In addition to their scientific merit, they are artistically ranked among the finest botanical paintings of any age. A number of Margaret Mee's finest paintings have been reproduced in two magnificent folio volumes: *Flowers of the Brazilian forests* and *Flowers of the Amazon*.[5] When reviewing her work for the Tryon Gallery opening in 1968, art critic and historian Wilfred Blunt said:

> *"They could stand without shame in the high company of such masters of the past, as Georg Dionysius Ehret and Redouté."*[6]

In writing the preface to Margaret Mee's diaries, published by Nonesuch Expeditions Ltd. in 1988, Professor Schultes notes:

> *"Of very special interest to me ... is her dedication to the forests of the Rio Negro area. It was here also that one of the greatest, but one of the least appreciated explorers of all time, the Yorkshire botanist Richard Spruce spent five productive years more than a century ago. In a very real way, Margaret honoured Spruce's extraordinary contribution. For the two have much in common: she is shy and self-effacing, as was he; and both, without actually realising the extent, have enriched our knowledge of the richest part of the Amazonian flora."*[7]

In this context, I would like to show how these two lives became integrally linked in history, despite the century that separated them. If one were to do a map overlay of Richard Spruce and Margaret Mee's journeys, one would be able to see the extent to which they explored the same regions. Although they both travelled extensively throughout the Amazon, it was the Rio Negro that appeared to evoke their greatest interest.

In his journals, Spruce described the river as follows:

> *"In many places, the river spreads out to an enormous width. Frequently it is sprinkled with islands, and sometimes opens out into a lake-like expanse, so wide that were it not for the lofty skirting forest, the opposite coast would be invisible."* (8)

For Margaret, it was the richness of its beauty and the secrets still held by its great forests that drew her on. Of Margaret Mee's fifteen journeys, nine were to the Rio Negro. As Professor Prance states so eloquently in the preface to her diaries:

> *"The Rio Negro region of Amazonia has called to the artist in Margaret Mee, with a call insistent and persistent. The ephemeral beauty, the light and shadow, the reflectance of the water, the magic of an opening flower bud have summoned her to ignore the frailties of body, and so to face repeatedly the rigours and dangers of river journeys of Sprucean extension."*(9)

On November 14, 1851, Spruce began his historic voyage up the Rio Negro. He had purchased a large canoe, capable of transporting up to six tons fully burdened and manned by six men. His crew proved to be excellent fishermen with the use of the bow, and fresh fish were regularly the order of the day. Spruce described his ascent up the Rio Negro as the:

> *"first agreeable voyage I have made in South America. The canoe being my own, I was master of my movements - could stop when I liked and go on when I liked. The cabin, too, was new and commodious. It was long enough to suspend my hammock within it, and I made myself besides a nice soft bed of thick layers of the bark of the Brazil-nut tree."*(10)

Both Spruce and Margaret Mee explored the Arquipelago das Anavilhanas on their way up the Rio Negro. This area is characterised by a maze of islands and igapós (naturally flooded forest) that extends to more than sixty miles upstream, with an estimated 500 islands. Margaret noted in her diaries:

> *"We were sailing amongst the islands, and trees that were gaunt and sculpted by nature stood in the water. Grass and floating plants formed a barricade between us and the forest, which was deeply flooded."*(11)

From this region, Margaret brought out a fine collection of plants that included the spectacular orchid *Cattleya violacea*, first discovered in the early 1800s on the Rio Orinoco by Alexander von Humboldt and Aimé Bonpland during their celebrated journey to South America. It was also collected by Spruce some fifty years later.

It was also in this region of the Rio Negro that Margaret finally found, after many years of searching, the night-flowering cactus, *Selenicereus wittii,* of which each blossom opens at night for *only* one night. It is believed to be pollinated by long-tongued hawkmoths. Near this area, a simple plaque has been erected as a memorial to her, which I saw in June 1993 when I visited that part of the Rio Negro.

On their respective voyages on the Rio Negro, Richard Spruce and Margaret Mee often encountered unpredictable weather. In his journals, Spruce noted: *"There is no foretelling the weather on the Rio Negro. When one looks for fair weather cometh rain, and the contrary."*[12] Margaret Mee's diary entry from her fifth journey seems to echo Spruce's words: *"The wind blew up suddenly as it does in a matter of seconds on the Negro – on other rivers, it never seems to break so quickly."*[13] On her eleventh journey, she described just such a scene:

> *"A storm was brewing as we reached the River Negro. Brilliant white clouds, piled high and masked by black, spread across the horizon. Then, quite suddenly, torrential rain without any wind broke over the glassy river. Towards dawn, a tempest swept over the forests with a terrific wind and pelting rain, and though the crew rushed to pull down the storm blinds I was drenched in my hammock."*[14]

Rapids or cataracts were another hazard of the upper Rio Negro which caused much distress to both Spruce and Mee alike, endangering weeks of hard labour by threatening to drench dry specimens and wash away cuttings and plantings so carefully preserved. Margaret Mee described these difficult passings as follows:

> *"...The boat had to be lifted over the rocks and dragged through the raging torrent. It took over an hour to pass the first fall, with much sweating and straining. The only way up was among the rocks near the bank and the two Indians used long poles to lever the hull across the obstruction which took until midday to pass."*[15]

Spruce described a similar situation on the upper Rio Negro at the perilous rapids just before reaching São Gabriel, a settlement which served as a collecting station for Spruce for several months during the years 1852 and 1853:

> *"I took the helm, though very ill-disposed for the task. The pilot leaped into the water with two or three more, applying their shoulders to the canoe, whilst the rest on board lugged at a rope made fast on shore beyond the point. In our course lay a sunken rock, which it was thought the canoe might pass; but instead, she struck on it and immediately fell over on one side. The boat swung around, forcing the rope out of the hands of the men, who instantly leaped into the water, and I was then left alone. I stuck pertinaciously to the helm. The canoe again swung round and fell over on the contrary side, and all thought this time she would have gone clean over; but she did not. Another revolution and she swung fairly off the rock, righting at the same time. I set her head to the fall and she shot down like an arrow."*[16]

Spruce considered his journey to São Gabriel a very successful one. During his two-month voyage, from November 14, 1851 to January 15, 1852, he had dried some 3,000 specimens, a much greater number than he had ever dried on any previous voyage.[17] He was particularly struck by *Heterostemon mimosoides*, which *"was in flower all the way up the river and formed a great ornament to its banks."*[18] This legume was later collected by Margaret Mee, and she completed her painting of it in 1978.

The very beautiful rain forest tree, *Gustavia pulchra*, was first discovered by Richard Spruce, and in a letter to a colleague in London dated April 15, 1852, Spruce noted that *"Gustavias were tolerably frequent* [on the upper Rio Negro], *but it was scarcely possible to preserve their flowers on account of the number of caterpillars bred in them."*[19] This plant was also collected and sketched by Margaret Mee in 1979.

Spruce's inhospitable accommodation at São Gabriel soon presented him with a host of new challenges. His house was infested with rats, vampire bats, scorpions, cockroaches and other pests of society. Spruce made special mention of two of these pests, the vampire bats and ants. His experiences were shared and confirmed by Margaret Mee over 100 years later.

Regarding the bats, Spruce observed that from the moment he entered his house for the first time, there were large patches of dried blood on the floor caused by the attacks of *"those midnight blood-letters"*[20] on former tenants. His first night in the house saw damage done to four toes of the inhabitants, followed by further nightly attacks against their finger ends, noses and chins. Margaret Mee also had her fair share of battles with these creatures of the night and sympathised more than once with Spruce's complaints. Her diary entry, written in August of 1967, reads as follows: *"In this region of the upper Negro, as well as in Saõ Gabriel de Cachóeira, vampire bats seem to be prolific and they are no better now than in the days of Richard Spruce."*[21]

In addition to these hardships, Spruce soon discovered, to his dismay, that food was extremely scarce in São Gabriel. He remarked about the trials caused by this shortage and confessed, at one point: *"Never was I so near dying of hunger."*[22] To make matters even worse, the saúba, or leafcutter ants, carried off a month's supply of Spruce's precious farinha. Even under these dire circumstances, his humour remained irrepressible, as illustrated by the following emphatic comment:

> *"They found my dried plants and began to cut them up and carry them off. I have burnt them, smoked them, drowned them, trod on them, and retaliated in every possible way, so that at this moment I believe not a saúba dare show its face inside the house."*[23]

The area surrounding São Gabriel is quite mountainous, and both Margaret Mee and Richard Spruce expressed attraction to these mysterious peaks, describing them as huge masses of granite rising abruptly out of the plains. Spruce sketched the Serra do Curicuriari from São Gabriel in 1852, a mountain which Margaret climbed on her third journey in 1964. Although Spruce did not climb this mountain, he did climb the Serra do Garna, a similar range close to São Gabriel. It was a disappointing expedition, as it involved arduous trail cutting and brought him little by way of botanical treasure. He described the mountain itself as *"being bar-*

ren of novelty", [24] with almost all the plants of interest being found in the caatinga forest at the base of the mountain. Margaret's ascent of the Serra do Curicuriari produced remarkably similar results, as she returned with just a *"few plants to treasure."* [25]

The main peak in the upper reaches of the Rio Negro is the Pico da Neblina, which Spruce sighted in the summer of 1853. Writing to Sir William Hooker, he noted: *"I could distinctly see, though at a great distance, the Serrania, called Pirápukú, or the long fish."* [26] Spruce never reached these mountains, and it was not until 1953, exactly 100 years later, that a team of botanists and explorers from the New York Botanical Garden, together with Venezuelan explorers, made the first recorded ascent of this mountain. The entire massif was named the Cerro da Neblina, 'Mountain of Mist', and it was confirmed that the mountain peak reached a height of 9,985 feet. [27]

Margaret became the first woman traveller to attempt the southern approach to this mountain. The expedition was financed by the National Geographic Society with the help of Dr Leonard Carmichael, and Margaret was accompanied by one of their photographers, Otis Imboden, who covered the story.

She began her climb in September of 1967. The 'Mountain of Mist' however, proved true to its name. The high level of precipitation had totally obliterated the path and it became impossible to reach the summit. A new trail would have taken weeks, if not months, to create. Margaret wrote, *"I was bitterly disappointed and could not restrain my tears"*. [28] The report of their expedition, with a list of species collected, including one new species, a bromeliad *Neoregelia leviana,* was published in the National Geographic Society Research Reports. [29]

The Rio Uaupés, a tributary of the upper Rio Negro, was an area extensively explored by both Richard Spruce and Margaret Mee, albeit a century apart. Spruce spent six and a half months along this river. He found the four Indian tribes represented *"among the finest of the South American tribes"*, [30] and persuaded several to sit for their portraits, including the Chief of the Tariana Indians. Spruce's simple drawings were a great success. He reported: *"The Indians were so delighted with the likeness of their chief that I verily believe every one came to have a look"*. [31]

Margaret spent almost two months in this region on her third journey in 1964. Both Mee and Spruce report finding a wealth of interesting plants. During this expedition, Margaret painted, among other plants, the beautiful bignonia, *Distictella magnolifolia*, which was first found by Humboldt in the early 1800s and then later by Koch in 1905, and the unusual plant, *Raputea paludosa*, which she described as *"an extraordinary swamp plant"*. [32] In fact, Margaret could scarcely contain her joy at finding such treasures and exclaimed, *"The whole area was a paradise of plants."* [33]

Spruce's uncanny ability to search out rare plants was evidenced very clearly during this time. In a letter to Sir William Hooker, dated June 27, 1853, he wrote that of his collection of some 500 species from this area, approximately four-fifths were as yet undescribed. It was here, on the banks of the Uaupés River, in 1852, that Spruce first discovered the lovely *Clusia grandifolia* which was subsequently collected and sketched by Margaret in this same area in 1983 (Fig. 2).

Both Spruce and Mee were disappointed to have to leave this region. Margaret felt that her time to leave came too soon and expressed a hope that one day she

Fig. 2. *Clusia grandifolia* (1983).

would return to explore this *"region of enchanting beauty"*.[34] Spruce was even more prolific in his praise, though it was laced with a touch of nostalgia:

> *"But when the time came for my return ... the weather cleared up, and, as we shot down among the rocks which there obstruct the course of the river, on a sunny morning, I well recollect how the banks of the river became clad with flowers, as it were by some sudden magic, and how I said to myself, as I scanned the lofty trees with wistful and disappointed eyes, 'there goes a new Dipteryx, there goes a new Qualea – there goes a new the Lord knows what', until I could no longer bear the sight, and covering up my face with my hands, I resigned myself to the sorrowful reflection that I must leave all these fine things 'to waste their sweetness on the desert air' ... and I have no doubt that [that] ... tract of country ... offers as rich a field for any botanist as any in South America."*[35]

The accomplishments of these two indefatigable explorers will continue to have significant impact on the future of Amazonia for many years to come. In his preface to Margaret Mee's diaries, Professor Schultes sums it up so succinctly:

> *"Their very different material contributions, her life-like paintings, his dried herbarium material of hundreds of species new to science, have given a powerful impetus to the growing outcry against the uncontrolled devastation of the largest rain forest left on the globe."*[36]

NOTES

(1)   L.B. Smith and M. Mee, *The Bromeliads*, South Brunswick, N.J.,1969.

(2)   M. Mee, *Margaret Mee – In search of flowers of the Amazon forests* (ed. T. Morrison), Woodbridge, 1988.

(3)   The Margaret Mee Amazon Trust, *The scholarship scheme, two years of achievement*, London, 1992.

(4)   S. Mayo, *Margaret Mee's Amazon*, Kew, 1988, 7.

(5)   M. Mee, *'Flowers of the Brazilian forests'*, London, 1968; M. Mee, *'Flowers of the Amazon'*, Rio de Janeiro, 1980.

(6)   W. Blunt, 'Book notes', *Journal of the Royal Horticultural Society* (1968), **93**, 307-8.

(7)   R.E. Schultes, in Mee, op. cit. (2), 12.

(8)   R. Spruce, *Notes of a botanist on the Amazon and Andes* (ed. A.R. Wallace), London, 1908, vol. 1, 276.

(9)   Mee, op. cit. (2), 11.

(10) Spruce, op. cit. (8), vol. 1, 269.

(11) Mee, op. cit. (2), 253.

(12) Spruce, op. cit. (8), vol. 1, 275.

(13) Mee, op. cit. (2),136.

(14) Ibid., 236.

(15) Ibid., 125.

(16) Spruce, op. cit. (8), vol. 1, 279.

(17) Ibid., vol. 1, 265.

(18) Ibid., vol. 1, 266.

(19) Ibid., vol. 1, 291.

(20) Ibid., vol. 1, 300.

(21) Mee, op.cit. (2), 122.

(22) Spruce, op. cit. (8), vol. 1, 298.

(23) Ibid., vol. 1, 293.

(24) Ibid., vol. 1, 297.

(25) Mee, op. cit. (2), 78.

(26) Spruce, op. cit. (8), vol. 1, 354.

(27) Mee, op. cit. (2), 98.

(28) Ibid., 118.

(29) M. Mee, 'Botanical studies in Northwestern Brazil' [in 1967], in *National Geographic Society Research Reports … of research and exploration authorised under grants from the …Society during the year 1966*, Washington, 1973, 147-156.

(30) Spruce, op. cit. (8), vol. 1, 328.

(31) Ibid., vol. 1, 326.

(32) Mee, op. cit. (2), 83.

(33) Ibid., 83.

(34) Ibid., 85.

(35) Spruce, op. cit. (8), vol. 2, 208-209.

(36) Mee, op. cit. (2), 12.

FURTHER REFERENCES CONSULTED

M. Goulding, *Amazon: the flooded forest*, New York, 1990.

V.W. von Hagan, *South America called them*, London, 1949.

G.T. Prance and T.E. Lovejoy (eds.), *Key environments: Amazonia*, Oxford, 1985.

C. Sandeman, 'Richard Spruce, portrait of a great Englishman', *Journal of the Royal Horticultural Society* (1949), **74**, 531-544.

R.E. Schultes, "Richard Spruce still lives', *Northern Gardener* (1953) **7**: 20-27, 55-61, 87-93, 121-125.

8

A contemporary botanist in the footsteps of Richard Spruce

Ghillean T. Prance

A contemporary botanist in the footsteps of Richard Spruce

Ghillean T. Prance
Royal Botanic Gardens, Kew, UK

Introduction

I have had the great privilege of three decades of exploration in the Amazon Basin and have visited many of the places where the great Yorkshire botanist Richard Spruce travelled; in so doing I have the greatest admiration for his work, and therefore feel it a great privilege to take part in the symposium celebrating his life. Rather than give my own account of the region today I want to draw attention to the acute powers of observation about so many things that are shown in Spruce's work. Therefore Spruce himself will talk to us from his *Notes of a botanist in the Amazon and Andes* edited by A.R. Wallace and published in 1908 (page references are given in the right hand margin below). My photographs show that much of what Spruce described so well can still be seen in the remoter parts of today's Amazonia. I will start at Belém where Spruce began his journey to Manaus, before travelling up the Rio Negro to Venezuela, and back to Manaus where the story will end with some observations about Amazonia today.

Spruce's botanical observations [1]

"Over these and other trees climbed Malphigiaceae, adorned with racemes of   5-6
yellow or pink flowers with elegantly fringed petals and usually a pair of large glands (or tubercles) at the base of each segment of the calyx; and still more showy Combretaceae, whereof one species (*Cacoucia* [= *Combretum*] *coccinea* Aubl.) was all in a flame with its long spikes of brilliant scarlet flowers.

Waste places, with a drier soil, were often clad with a vigorous but weedy vegetation, the predominant plants being rank prickly Solana, with large woolly leaves and apple-like fruits, and several species of *Cassia*, gay with golden flowers, which were followed by long pods whose loose seeds kept up a continual rattling as one pushed through the interwoven branches."

"At Tauaú I first realised my idea of a primeval forest. There were enormous   17
trees, crowned with magnificent foliage, decked with fantastic parasites, and hung over with lianas, which varied in thickness from slender threads to huge python-like masses, were now round, now flattened, now knotted, and now twisted with the regularity of a cable. Intermixed with the trees, and often equal to them in altitude, grew noble palms; while other and far lovelier species of the same family, but bearing plume-like fronds and pendulous bunches of black or red berries, quite like those of their loftier allies, formed, along with shrubs and arbuscles of many types, a bushy undergrowth, not usually very dense or difficult to penetrate."

"The noblest trees in the forests of Tauaú were the Bertholletiae, and one [18] specimen was perhaps as large a tree as I have anywhere seen in the Amazon valley. Its nearly cylindrical trunk, not at all dilated at the base, measured 42 feet in circumference, and at 50 feet from the ground it seemed almost fully as thick. It began to branch at about 100 feet, so that its crown rose high above the surrounding trees, but I could not see it distinctly enough to be able to form an idea of the entire height. I suppose the Bertholletiae and Eriodendra (Silk-cotton trees, in Lingoa Geral, Samaúma) to be the loftiest trees in the Amazon valley...."

"... Almost the first thing that strikes the observer is the enormous dilation at [20] the base of many of the trunks, in the shape of broad, flat, subtending buttresses, more or less triangular in outline, and rarely exceeding 6 inches in thickness, set around each trunk to the number of from four to ten. These buttresses are really exserted roots, or, as the Indians correctly call them sapopemas (*sápo*, a root; *péma*, flat)...."

"... Not infrequently they are fantastically twisted, and the outer edge may [21] be either straight or bulged outwards; but in all cases their woody fibre is in a state of extreme tension, so that on striking an axe or cutlass suddenly into them they give out a sound like the breaking of a harp string. On examining attentively trees which have sapopemas notably developed, it will be found that they have no central or tap root at all, nor do the lateral roots dip deep under the soil."

Fig. l. A three-toed sloth.

Fig. 2. A rainforest view.

Fig. 3. An Amazonian species of *Aristolochia*.

Fig. 4. Escada de Jaboti or the monkey ladder vine (*Bauhinia* sp.).

"In Moraceae, especially in the parasitic (or, properly speaking, epiphytal) fig   24
trees, we have another type of sapopemas, whose origin is plain enough. The
excrement of a bird, containing seeds of figs on which it has fed, falls on the fork
of a tree, or even on the bare trunk or branches, to which it adheres; there a seed
germinates, and as its stem grows upwards, its root, in the form of a broad plate
- soon enlarging into a sheath, if the mother tree be slender - pushes down-
wards, diverging a little from the vertical on all sides, and dividing into a number
of forks, seeks the ground. If the height be great, the forking is repeated several
times, giving the appearance of so many pairs of marauding legs descending
from the upper part of a habitation, to which they had gained access one does
not at first see how, and feeling for the ground with their toes. Having reached
the ground, they plunge therein, increase rapidly in breadth, by the addition of
matter to their outer edge, but scarcely at all in thickness, so as to form plank-
like buttresses, and the parasite having thus gained an independent footing,
straddles over the too often lifeless trunk of the friend whom he has crushed to
death in his embrace, when his support is no longer needed."

"Of all lianas, rope-plants, or sipós (as they are called in Tupi), the most fan-   28
tastic are the Yabotím-mitá-mitá or Land-turtle's ladders, which have
compressed, ribbon-like stems, wavy as if they had been moulded out of paste,
and while still soft indented at every few inches by pressing in the fist. They are
usually not more than three or four inches broad, but I have sometimes seen
them as much as 12 inches: and they reach two or even three hundred feet in
length, climbing in the tree-tops, passing from one tree to another, and often
descending again to the ground. They belong to Schnella [= *Bauhinia*], a genus of
Leguminosae, and are found all through the Amazon valley. The commonest
species near Pará is *Schnella* [= *Bauhinia*] *splendens*, Benth."

"Birthworts (Aristolochiae) are notable for their thick bark, cloven down to   29
the woody axis in six or more furrows. When cut across they give out a strong
smell, usually rather fetid, but in some cases pleasantly aromatic. They are scarce
in the plains of the Amazon valley, and their singular hooded and often lurid-
coloured flowers are difficult to find."

"... The Indians profess to know several lianas whose juice affords a copious   31
and wholesome draught, but I could never trust myself to drink of any but the
Dilleniaceae, chiefly of the genus Doliocarpus. For this purpose it is not sufficient
merely to sever the liana, when only a small quantity of fluid would gush out...."

"And now a word about the flowers. Were a naturalist to combine into one   40-41
glowing description all the gay flowers, butterflies, and birds he had observed in
any part of the Amazon valley, during a whole year, he might no doubt produce a
most fascinating picture, which would, however, utterly mislead his readers, if
they were thereby led to suppose that even a tithe of these beautiful objects were
ever to be seen together, or in the space of a single day. Very much depends on
seeing any particular site exactly at the time when its most showy plants, insects
or birds are in greatest perfection of profusion; and the effect is always modified
by the peculiar tastes of the observer. To the naturalist, the mere fact of an object's
being new and strange invests it with a conventional beauty, independent of all
aesthetic considerations; and for myself I must confess that, although a passion-
ate admirer of beauty of form and colour, and with a most sensual relish of

Fig. 5. The water vine (*Doliocarpus* sp.).   **Fig. 6.** A species of *Heliconia.*

exquisite odours, I recall with the greatest zest those scenes which yielded me the greatest amount of novelty. But we are again losing sight of the flowers, and, sooth to say, the flowers of Amazonian trees are often so inconspicuous, either from their minuteness or from their green colour assimilating them to the leaves, that none but a botanist ever would see them. There are doubtless many glorious exceptions; but it was not until some years after I had left Pará, and had penetrated to the northern border of the Amazon valley, that I realised my preconceived notion of the loftiest trees of the forest bearing the most gorgeous flowers. At Pará the Leguminifers and Bignoniads, both as trees and as lianas, outshine all other orders in the abundance and beauty of their flowers."

"... Most novel to the European botanist are the curious leathery, dull-coloured, [41] but often richly-scented flowers of the Anonads or Sour-sop family; the large white or roseate flowers of the Lecythids or Monkey-cups, notable for the stamens being borne on a large hooded receptacle in the centre of the flower...."

"... Of all families of plants - excepting perhaps Leguminifers - Rubiads seem [42] to occupy the principal place in the Amazon valley, from the shores of the Atlantic to the crests of the Andes. They are always easily recognisable by their opposite entire leaves, with interposed stipules; and by their tubular flowers. The latter are often of extraordinary beauty, and coupled with the great importance to man of the products of many rubiads - for where else do we get stimulants so precious as coffee and quinine? - render these plants surpassingly interesting to the traveller...."

"... There congregate the Heliconiae, looking like their near allies the Musae   46
or plantains, but their flower-spikes garnished with showy scarlet bracts; various
species of Maranta, Alpinia, Thalia, etc., all having foliage approaching that of the
Cannae now so much cultivated in our gardens; two or three species of Costus,
looking like gigantic spiderworts, etc."

"Nor must we omit to mention the roots that creep and cross each other   47
everywhere along the ground, or rise above it in buttresses, arches, or loops,
which must be climbed over or under; nor the huge, rotting, reeking trunks -
corpses of fallen giants of the forest - partly overrun with mosses, ferns, and
lianas. Sometimes a prostrate trunk appears still sound - even the bark is entire
- yet it has already been excavated by the voracious termite, so that it yields
with a crash when stepped upon, probably prostrating the traveller, and not
infrequently disturbing the repose of the snake or toad which has taken up its
abode in the cavity."

"... At the bottom was a grove of palms. chiefly of two species, the before-   48
mentioned Assaí and the Paxiuba (*Iriartea exorrhiza*). The latter most singular
palm has the trunk supported on, not a tripod, but a polypod, of exserted roots -
the spokes of a half-spread umbrella may give a very good idea of them, suppos-
ing a few additional spokes to be inserted between the circumferences and the
axis. Each root or spoke is a rigid cylinder, some two inches in diameter, so beset
with hard prickles that it may and often does serve as a grater."

Spruce's economic botany observations [1]

"I ought not to take leave of Pará without adding a few words on the products   50
of the forest that enter so largely into the consumption and commerce of that
port. A complete account of their economic and medicinal uses would, however,
require a separate volume; and as many of them, such, for instance, as sarsaparilla, are collected in the far interior, and are only taken down to Pará for sale and
for re-embarkation to Europe and North America...."

"On the 24th of August we visited an Indian settlement by an igarapé, about   10
five miles inland from Mr Campbell's house, in order to see the manufacture of
fireproof pottery, and especially the Caraipé tree, in whose bark (mixed with the
clay) was said to reside the fire-resisting property...."

"One of Mr Campbell's mulattos accompanied me as a guide. Leaving the   10-14
beaten track, he took us by a short cut through the forest, along a hunter's trail,
where my unpracticed eyes could scarcely distinguish any semblance of a path.
We reached the igarapé, which was not very wide, but as such conveniences as
bridges were almost unknown in that region, we should not have been able to
get across if our guide had not swum over and brought us a canoe from the
other side. A few steps beyond stood the four or five cottages we were in quest
of, embosomed in a grove of orange trees and plantations. I surveyed them
with interest, for they were the first abodes of the dwellers of the forest I had
seen, although there were some of mongrel character (like their inhabitants) in
the Nazaré and other suburbs of Pará. They wore an air of neatness and com-
fort, and made me think of Will Atkin's house on Robinson Crusoe's island.
The walls were of palm leaves, closely woven into a sort of matting. The roofs

The process of making pottery with the addition of caripé (*Licania octandra*) bark ashes to harden the pots as described by Richard Spruce and photographed here on Marajó Island in 1987.

Fig. 7. A bowl of caripé ash.

Fig. 8. Balls of clay being covered with caripé ash.

Fig. 9. Close-up of a ball of clay which has been rolled in caripé ash.

Fig. 10. The firing is done individually with the fire built around the pot.

Fig. 11. The bowl in front has been glazed with the resin of jatobá (*Hymenaea courbaril*).

Fig. 12. The potter is holding a forked stick with a piece of resin which is applied to the hot bowl to form the glaze.

Fig. 13. A Marajó water pot made from clay mixed with caripé.

were covered with a sort of shingles, made by tying several of the broad flat fronds of a small palm called Ubím (*Geonoma*) on to a stick so as to closely overlap each other. A roof of Ubím looks pretty, keeps out the rain well, and lasts a long time. At a short distance was the essential mandiocca plantation, covering several acres. An old Indian pointed out to me eight or nine varieties of that most useful vegetable (*Manihot utilissima* of botanists), each grown in a plot kept carefully separate from the rest; he professed to distinguish them by the leaves, but I confess I was unable to do so; however, there is no doubt that the roots vary much in shape and colour, some being whitish, others deep yellow; that some kinds ripen sooner than others, and that some suit best for making farinha de agua, others for farinha secca. Farinha de agua is made by macerating the mandiocca roots in water till they are soft enough to be broken up by hand. Farinha secca is made entirely from the fresh grated roots. The former contains nearly all the starch in combination with the other nutritive constituents, but the latter has parted with most of the starch in the repeated washings and squeezings the pulp undergoes to free it from the poisonous juice. When the main object is to have the tapioca or mandiocca starch separate, the pulp of the grated root is alone employed.

"I was then shown the Caraipé pottery, which comprised almost every kind of cooking utensil. It was made of equal parts of a fine clay, found in the beds of igarapés, and of calcined Caraipé bark; but in other places where I have seen the manufacture carried on (and there is no Indian's house in the Amazon valley where it is not familiar) a much smaller proportion of the bark was used. The property which renders the bark available for this purpose is the great quantity of silex contained in it. In the best sorts - such as I afterwards saw on the river Uaupés - the crystals of silex may be observed with a lens even in the fresh bark; and the burnt bark turns out a flinty mass (with a very slight residuum of light ash, which may be blown away), so that for mixing with clay it requires to be reduced to powder with a pestle and mortar. The bark I saw at Caripi is, however, much less siliceous, and when burnt may be broken up with the fingers.

"Having satisfied my curiosity as to the pottery, we started into the wood to see the Caraipé tree, and after much searching found one - a straight slender tree whose height I estimated at 100 feet; and it was branched only near the summit, so that it was impossible from below to say what the leaves were like. A young Indian offered to procure them for me, and then I witnessed for the first time the Indian mode of climbing any tree not of inordinate thickness. A handkerchief is tied by the two opposite corners, or a bit of rope about 2 feet long by the two ends, or, better still, because everywhere obtainable in the forest, a ring of sipó is made of the same size. The climber, standing at the foot of the tree, puts the toes of each foot into the ring and stretches it to its full extent; then, embracing the tree with his arms - or grasping it with his hands if it be very slender - he draws up his legs as far as he can, and holding the ring tight to the tree with his feet, so as to form a sort of step, he straightens himself out and repeats the process; so that by a series of snail-like movements (I mean as to the attitudes, not the pace), he soon reaches the top of the tree ...."

"The Indian brought down branches of the Caraipé, but they unfortunately possessed only leaves, no flowers or fruits. Defective as they were, my dried specimens were placed in the hands of Mr. Bentham, and his vast knowledge of what may be called comparative vegetable anatomy enabled him to assign them, nearly with certainty, to the order Chrysobalaneae, and even to indicate the genus (*Licania*) to which they probably belonged. I afterwards fell in with several sorts of Caraipé trees, and was fortunate enough to gather flowers and fruits of some of them, which confirmed Mr. Bentham's opinion of their being species of Licania. The leaves are mostly like those of our apple and pear trees, although the Licaniae are in reality more nearly related to the plum tribe (Drupaceae), and the small sub-globose drupes are not unlike very small and prematurely-ripened peaches in their downy skin, usually painted on one side with carmine or purple, but they are very dry and scarcely edible."

"... I had also two Mirití palms cut down, for the sake of truncheons of their trunks to send to the Museum of Vegetable Products at Kew. There were two forms considered distinct species by Von Martius, viz. *M. flexuosa*, which has the fruits nearly globose; and *M. vinifera*...."

Voyage to Santarém

"We embarked in the *Tres de Junho* on the 10th of October, at 9 pm. Our 55-56
course lay at first westerly, trending a little south, across the bays of Marajó and
Limoeiro ... Still keeping the isle of Marajó on our right, we entered a narrow
channel called the Furo dos Breves, on which stands a small village of Breves.
Our course began now to trend a little northerly, and after crossing a deep lake
called the Poço (well), we entered another channel (Canal de Tagipurú) which,
after a long winding course, brought us finally into the Amazon.

The Poço was a great rendezvous of floating aquatics, detachments of which
made excursions a little way up the Tagiupurú with the flood-tide, then back
again and a little way down the Furo dos Breves with the ebb-tide. The Tapuyas
called them all Mururé, but they were made up of plants of widely distinct families, the most abundant being the common *Pistia Stratiotes*. ... Another Mururé
was the singular *Pontederia crassipes*, [= *Eichhornia crassipes*] which bore short
spikes of pale blue flowers springing from among the roundish leaves, whose
stalks became inflated and filled with air, so as to serve as floats."

"... The islands were mostly densely wooded; but one of them (probably of 60
recent formation) presented the appearance of a beautiful meadow, being clad
with long grass, sprinkled with low trees with here and there a clump of arbores-
cent Aroids; and begirt by a natural fence of *Salix Humboldtiana*, a graceful
willow, notable for its long, narrow, yellow-green leaves, and for its being distributed, in varying forms, along the banks of rivers of *white* water (but not of *black*)
throughout equatorial America ..."

"The local name of the Tapajoz at Santarem is "Rio Preto" or Black River, but 64
the real colour of its waters is a deep blue. When I first saw it, in the dry season,
the blue water extended down the southern side of the river for several miles
below Santarém, before being absorbed in the muddy expanse of the Amazon ..."

Fig. 14. Floating mass of water hyacinth (*Eichhornia crassipes*).

The Savannas of Santarém

"Instead of the forest-clad plains and artificial pastures of Pará, I found at   65
Santarem natural campos or savannahs, sloping gently upwards from the banks
of the Tapajoz, and at the back rising into picturesque but not lofty hills...."

"The vegetation of the upland campos reminded me of an English pleasure-   66-67
ground. It consisted of scattered low trees, rarely exceeding 30 feet in height,
and here and there beds of gaily-flowering shrubs, with intervening grassy
patches and lawns. The grass in the dry season looked rather dreary, for it consisted of but one species of Papsalum, growing (like many tropical grasses) in
scattered tufts, whose culms and bristle-like leaves were hoary with white hairs;
so that it differed widely from the dense green turf of an English meadow.
Among the trees then in flower, the Cajú or Cashew-nut (*Anacardium occiden-
tale,* L.) was exceedingly abundant; and an old Cajú, with its rough bark, its
branches touching the ground on every side, its young leaves of a delicate red-
brown, and its numerous pear-like yellow or red fruits (more properly enlarged
fruit-stalks), each tipped with a kidney-shaped knob (the real fruit), is a picturesque object, notwithstanding its humble size. With the Cajú grew the
Caimbé (*Curatella americana*, L.), a small tree not unlike a stunted oak in habit
and in the sinuated leaves, which are, however, so rough that they are used in
lieu of sandpaper by the carpenters of Santarem."

Victoria amazonica

"... During my voyage from Pará I had learnt from the Tapuyas that in lakes   75-76
around Santarem there was a water-plant called in Portuguese the Forno or
Oven, in Lingoa Geral Auapé-yapóna (the Jacaná's oven), from the resem-
blance of its enormous leaves to the circular oven used for baking farinha, and
from the little river-side birds called Jacaná or Auapé being frequently seen
upon them. Captain Hislop and other residents at Santarem confirmed this
report, which pointed plainly to the Victoria. Having obtained precise directions to one of its localities, Mr Jeffries was so kind as to lend me a boat and
men, and to accompany myself and Mr Wallace to see the Forno. We crossed
the main channel of the Amazon to what appears from Santarem to be its
northern shore, but is really the north side of a very long island, called Ananarí;
and then went a little way up a creek to a sitio called Tapiírauarí. A walk thence
of about two miles across the island brought us to a paraná-mirí, in which we
had the satisfaction of finding a patch of the Victoria about 10 yards in diameter. There was barely 2 feet ot water where it grew, rooted into nearly an equal
depth of mud. The leaves were packed as close as they could lie, and none of
them exceeded $4^{1}/_{2}$ feet in diameter. I wished to obtain proof as to whether its
duration was annual or perennial, but was unable to decide, although the evidence seemed in favour of the latter. I found no prostrate submerged trunk, but
a thick central root penetrating so deep that we could not dig to the bottom of
it with our terçados. This root, notwithstanding its size, might be annual; but
then every one who knew the plant assured me that the Forno was never
wanting all the year round in that and other localities; in which I afterwards
found them to be correct; not so, however, in their statement that, when the

Fig. 15. The royal water lily (*Victoria amazonica*).

lakes and creeks rose to their winter level, not only did the petioles lengthen out to keep pace with the rising waters, but the floating leaves went on increasing proportionately in diameter, until they sometimes attained a breadth of 12 feet. I found, in this and other instances, that the measuring tape was needed to correct the illusions caused by the exaggerated statements of others, or even by the apparent evidence of my own senses.

The Water-lilies I have since seen in South America are certainly all of them annual; and one which springs up on the savannahs of Guayaquil, when the winter rains transform them into lakes, takes only from two to three months to attain its full dimensions and ripen its edible seeds."

Spruce's botanical observations [2]

"Damp shady hollows, where the vegetable mould lay deep, were often over- 99
spread with *Helosis brasiliensis*, Mart. (of the natural order Balanophoraceae), one of the lowest forms of flowering plants, looking quite like the young state of some fungus (Agaricus or Polyporus), until what seems to be an unexpanded cap is found to be a solid oval head of a reddish-brown colour, studded with minute flowers of the most rudimentary structure. I have seen it at several points in the Amazon valley, and it reappears near the coast of the Pacific, at the western foot of the Andes."

"... These floating Grass-islands are a sure indication of the river beginning 109
to rise, and they merit a particular description here, from being a remarkable

105

and indeed unique characteristic of the Amazon and of its tributaries with white or turbid water, but not of those with blue or black water, nor indeed of any other rivers in the world that I have seen or read about. The rafts of driftwood on the Orinoco, described by Humboldt, and seen there most lately by myself, have their counterparts on the Amazon, the Mississippi, etc.; but the Grass-islands of the Amazon are totally different things; they are compact masses of grass, in a growing state, varying from 50 yards in diameter to an extent of several acres. What kind of grass they consist of, and how they came there, I will now try to show.

Along low shores of the Amazon, especially in deep sheltered bays, there is often a broad belt of Caapím (the Tupi name for grass, in general); and the same feature, more strongly marked, is seen in some of the still paraná-mirís, and in lakes that communicate with the river by a short channel. This Caapím consists chiefly of two species, the Canna-rana or Bastard-cane (*Echinochloae* sp.) and the Piri-membéca or Brittle-grass (*Paspalum pyramidale*) - amphibious grasses, for whose production *white* water is essential, as is proved by their absence from the Tapajoz and Rio Negro throughout their entire course, and free from the Trombetas above the Furo de Sapuquá."

"... It was strange, also, to see great quantities of a floating Sensitive-plant, 115 *Neptunia oleracea*, whose slender tubular stems were coated with cottony felt of an inch in thickness, as buoyant as cork, serving to sustain completely out of water the heads of pale yellow flowers, and the delicate bipinnate leaves, which shrank up at our approach."

"Among the very few palms at Santarem, one, the Jará (*Leopoldinia pulchra*, 150 Mart.), grows gregariously by the Tapajoz; and it reappears on the Rio Negro in such abundance as to be one of the characteristic plants of that river. It is of humble growth, rarely exceeding 12 to 15 feet, and its most marked feature is the rigid leaf-sheaths, split into finger-like divisions, which remain clasping the stem like so many gauntlets after the leaves themselves have fallen away."

"... But more abundant than any of these, and (as I afterwards found) extend- 153 ing along the banks of the Amazon to the very roots of the Andes, was the Pao Mulatto or Mulatto tree [= *Calycophyllum* (Rubiaceae)], so called from the colour of its bark, which is continually peeling off and being renewed. It grows 60 to 100 feet high, and branches in such narrow forks that its top is usually in the form of a reversed cone....".

Spruce's economic botany observations [2]

"... Copaifera from which capivi is obtained in great quantities along various 162 tributary streams of the Amazon. All the species have the small flowers closely set on the branches of a rigid pinnate panicle, the flattened pink ovary standing out beyond the four or five white petals and the free stamens (eight to eleven); and the leaves consist of two or more pairs of deep green leaflets beset with pellucid dots. In old trees the trunk becomes hollow at the core, and there the oil accumulates and is extracted by boring with an auger....".

"Pitómba (*Sapindus cerasinus*, sp. n.), a shrub 6 to 10 feet high, with pinnate leaves and white flowers, grows on stony slopes, at Cape Mapirí and elsewhere,

Fig. 16. The fruit of pitomba (*Talisia* sp.).

on the Tapajoz. It bears a yellow fruit the size of a cherry, and has something of the same taste. The thin pulp envelops a single seed, which, on tasting, I found to have a pleasant flavour of black currants, and therefore ate several of them: nor did any ill consequences result, but when I told my Santarem friends of it, they had never known of the seeds being eaten, and that I had acted imprudently, for the plant belonged to a poisonous family. I knew, however, that the seeds of the nearly-allied Guaraná were wholesome, and I afterwards found the seeds of most of the Sapindaceae are at least harmless, notwithstanding the deadly properties of the stems and roots of such plants as *Paullinia pinnata*.

"Yenipápa (*Genipa macrophylla*, sp. n. - Cinchonaceae - and other two new species) - *Genipa americana*, L., the most widely distributed species of the genus, I have seen wild in many places across the whole breadth of South America. In Peru it is called Huítu; in Ecuador, Jagua. Its fruit (a large olive-green berry) affords a permanent black dye, and is in universal use by the Indians for staining their skins; it is also pleasant eating, when allowed to become over-ripe, having then the consistence and much of the flavour of the medlar." 164

"The Guaraná plant (*Paullinia Cupana*, Humb. and Bonpl., of the natural order Sapindaceae) is a stout twiner, whose scandent propensities are kept down in cultivation, so as to reduce it to a compact bush with sinuous entangled branches. The leaves are pinnate, of five leaflets, each nearly half a foot long, oval, and coarsely serrated. The racemes have small white flowers set on them in clusters, and in fruit are pendulous. The fruits are about an inch and a half long, pear-shaped, with a short beak, yellow, passing to red at the point; and they enclose a 180-181

Fig. 17. A rubber gatherer's dwelling has not changed much since the days of Richard Spruce.

Fig. 18. Coagulating rubber with the use of acid smoke.

single black shining seed about three-quarters of an inch in diameter, half-enveloped in a white cup-shaped aril.

The fruit is gathered when fully ripe, and the seeds are picked out of the pericarp and aril, which dye the hands of those who perform the operation a permanent yellow. The seeds are then roasted, pounded, and made up into sticks, much in the same way as chocolate, which they somewhat resembled in colour. In 1850 a stick of guaraná used to weigh from one to two pounds, and was sold at about one milreis (=2s. 4d.) the pound at Santarem; but at Cuyabá, the centre of the gold and diamond region, it was worth six or eight times as much. The usual form of the sticks was long oval or subcylindrical.

... The intense bitterness of the fresh seed is dissipated by roasting to a much greater extent than it is in coffee, and a slight aroma is acquired. The essential ingredient of guaraná, as we learn from the investigations of Von Martius and his brother Theodore, is a principle which they have called guaranine, almost identical in its elements with theine and caffeine, and possessing nearly the same properties. Guaraná is prepared for drinking by merely grating a small portion - say a tablespoonful - into cold water, and adding an equal quantity of sugar. It has a slight but peculiar and rather pleasant taste, and its properties are much the same as those of tea and coffee, being slightly astringent, and highly stimulating to the nervous system. It has had the reputation of a powerful remedy against diarrhoea, but I never found it so, although I have tried it largely, both on myself and other people. The general notion, however, is that guaraná is a preventive of every kind of sickness, and especially of epidemics, rather than an antidote against any; and Martius says of it "pro panacea peregrinantium habetur". Its immoderate use relaxes the stomach and causes sleeplessness - and precisely the same effects as result from the abuse of tea and coffee."

"... A stout liana is wound round the trunk of each Seringa tree, beginning at the base and extending upwards about as high as a man can reach, and making in this space two or three turns. It supports a narrow channel made of clay, down which the milk flows as it distils from the wounded bark, and is received into a small calabash deposited at the base. Early in the morning a man starts off into the forest, taking with him a terçado and a large calabash (called a cuyamboca) suspended by a liana handle so as to form a sort of pail, and visits in succession every Seringa tree. With his terçado he makes sundry slight gashes in the bark of each tree, and returning to the same in about an hour he finds a quantity of milk in the calabash at its foot, which he transfers to his cuyamboca. The milk being collected and put into large shallow earthenware pans, other operators have meanwhile been filling tall, narrow-mouthed Caraipé pots with the fruits of the Urucurí palm and setting them over brisk fires. The smoke arising from the heated Urucurí is very dense and white; and as each successive coating is applied to the mould - which is done by pouring the milk over it, and not by dipping it into the milk - the operator holds it in the smoke, which hardens the milk in a few moments." 185

"The change from the yellow water of the Amazon to the black water of the Rio Negro is very perceptible, and indeed abrupt. The latter is black as ink when viewed from above, and stones or sticks at the bottom seem red; but when taken up into a glass it is a pale amber colour, and quite free from any admixture of mud." 200

"On the Jauauarí I saw a small plantation of Ipadú, a shrub of which the pow- dered leaves are chewed by the Indians throughout the Rio Negro. I found it to be (as I had expected) the *Erythroxylon Coca*. The leaves are roasted and then pounded in a mortar made of the trunk of the Pupunha palm, from 4 to 6 feet long, the root being left on for the bottom and the soft inside scooped out. It is made so long on account of the impalpable nature of the powder, which would otherwise fly up and choke the operator; and it is buried deep enough in the ground to be worked with ease. The pestle is made of any hard wood. When sufficiently pounded they are mixed with a little tapioca to give it consistency. With a chew of Ipadú in his cheek an Indian will go two or three days without food, and without feeling any desire to sleep."

Fig. 19. Ipadu leaves (*Erythroxylum coca* variety *ipadu*) being dried for pulverising and mixing with food by a Maku Indian.

A white sand campina

"There is another campo near the Barra, on the same side of the river, which 218-219 differs much in every respect from the one I have described above. It is elevated about 100 feet above the river, and the soil is a loose white sand. The vegetation is chiefly shrubby, and one shrub called Umiri is so abundant that the campo is called from it the "Umirisal". It is a species of Humirium belonging to a small natural order (Humiriaceae) peculiar to tropical America, and bears a fruit which is said to be very agreeable ... But what rendered the campo most interesting in my eyes was that here and there on the burning sand were large patches of four species of Claydonia [*Cladonia*], two of them exceedingly like our common Reindeer Moss, and a third with bright red fruit looking quite like our *C. coccinea* [*C. coccifera*]. When I add to this that everywhere among the bushes grew up a tall Fern (*Pteris caudata*) scarcely distinguishable from our Common Brake, it will easily be seen how strongly I was reminded of an English heath. There were, however, two Ferns of the curious genus Schizaea - one preferring the most exposed situations, the other nestling under bushes, and both in considerable quantity - looking so very *tropical* as at once to disperse the illusion, if it had entered my head to fancy myself at home."

[I might add here that I only travelled on the Rio Negro because on a trip on the Rio Purus I nearly died of malaria and my Brazilian doctor ordered me to travel where there were no mosquitoes so as not to get reinfected. We shall see what happened to Spruce in that region. - GP]

Fig. 20. An Amazonian campina or white sand forest.

Spruce's botanical observations [3]

"My Rio Negro collections include samples of nearly every natural order of 220
plants. Leguminosae continue to constitute a large proportion of them, but
Caesalpiniae and Mimoseae are more numerous then Papilionaceae, which was
not the case in the localities previously visited. I have several large-flowered
Loranthi not found at Santarem, numerous Rubiaceae, Myrtles, and Melastomas
almost without end, and some curious intermediate forms between these two
orders. Lecythideae are not scarce, but many of them very difficult to access on
account of their large size. The small-fruited species of Lecythis are called by the
Indians Macacarecuya or the Monkey's drinking-cup, their fruit quite resembling
a cup, when the lid has fallen off."

"A palm much cultivated in the Barra and the adjacent sitios, and said to 223
grow wild up the Rio Negro, is the Pupunha, which I suppose to be the same as
the Pirÿäô (*Guilielma speciosa*, Mart.) mentioned by Humboldt as growing on
the Upper Orinoco. The fruit of this is perhaps more valuable as an edible than
any other palm-fruit; the sarcocarp contains a large quantity of starch, and it is
sometimes developed to such a degree that the nucleus is quite obliterated.
Eaten with salt, the boiled or roasted fruit much resembles a potato, but it is
also very pleasant eating with molasses. A spadix of Pupunha, laden with ripe
fruit, is one of the most beautiful sights the vegetable world can show: the fruits
are of the clearest scarlet in the upper half, passing below into yellow, and at the
very base to green."

"I enclose you two flowers of a Leguminous tree which was in flower all the 266
way up the river and formed a great ornament to its bank. It is a Heterostemon
(a most remarkable genus), but whether a described species I cannot say. The
petals are a fine blue slightly tinged with purple, and the column of stamens is
red. There are no pods ripe yet, but I will try to send you some. As it often flowers at 10 feet high, it is very suitable for cultivation.

"... there are, however, plenty of Podostemons on the granite rocks which 267
peep out of the river (and, by the by, make the navigation very dangerous), but
all, *all* dead and burned up. It is here, as I remarked at Santarem, the
Podostemons all flower just as the water leaves them, that is, early in the dry season; and my ascent of the Rio Negro was made towards the close of the dry
season; but if I live, these little fellows shall not escape me."

Serra Curicuriarí

"Jan. 10. - This morning at 8, Senhor Pailhête took me across the river to view 280
the Serras de Curicuriarí, which lie directly at the back of his sitio (a day's journey,
but there is no path), and on the east side of the river Curicuriarí ... From our
point of view they might have been clearly seen had there not been much vapour
in the air. The highest has much steep rock, mottled with brown and white, and
quite inaccessible on the south side, but its summit might possibly be reached by
taking a col between it and the flat-backed wooded mountain to the right."

"We went on continually aiming for the highest ground, as well as the blocks 309-31
of granite and network of sipós would allow us. We struggled on, sometimes
climbing steep inclined planes of slippery stones by the aid of the sipós and roots

on them, until we both began to feel rest needful. We sat down, and opening the barometer I found we had already climbed 1000 feet. I felt sure, therefore, that we had already reached above half-way up, and I bade my guide take courage. Our companions here joined us and we resumed our march. In a short time we emerged on a narrow ridge which sloped rapidly down on the opposite side, and we correctly judged it to be a shoulder of the mountain connected with the terminal peak ... During this ascent of the peak we were in the midst of a thick cloud and were soaked by the wet dripping from the trees. Though we cut a way at the summit to the side from which we should have had a good view of the rest of the serra and of the river, and waited some time, the clouds only now and then partially rolled away so as to show the first lower ridge round the base of which we had skirted in order to reach the foot of the highest peak. It seemed to be continuous with the latter, being joined by the shoulder before mentioned and forming with it a kind of cirque. We were on the top exactly at noon."

The Rio Negro

"... We had one considerable fall to ascend just after starting, but after this we had only rapids easily passed until reaching the worst of all the falls, at the foot of the hill on which Saõ Gabriel is built. It is commonly called the "cachoeira de praya granda" from a wide sandy beach stretching below it, on the left bank of the river. Here we had again to pass the heavy cargo overland." 286

"I sat down under a cliff of granite, watching with anxious eyes the passage of my little vessel; and when at last she had plainly cleared the perilous spot, a load was, as it were, removed from my heart, and I mentally returned thanks to a kind providence who had thus brought me safely through all the dangers of the voyage, and had permitted me to reach its termination without losing either my vessel or a single article of her cargo, the latter to me invaluable." 286-287

"Like the Rio Negro and Solimoẽs, the Uaupés is said to be at its height near June 24, but does not fall perceptibly till the beginning of August. When I reached São Jeronymo on September 7, it was gradually lowering, and so continued, only occasionally filling again a few inches with a heavy rain. But on the 20th of November it began to refill, and by midnight had risen 20 inches. Afterwards it rose very slowly until December 5, when it again began to fall, the whole rise not having exceeded 3 or 4 feet. 332-333

It went on falling a few inches each day, or on some days neither rising nor falling, till December 19, when it began to rise again, and by the 23rd had reached the height of its former rise. Thus it continued (save that on one day, the 28th, it fell a little) until midnight on the 31st, when it began to subside ... then about midnight it began to rise rapidly, and continued rising until February 15, having reached within a foot of its former rise."

"Don Diego is perhaps the only white now living in the Canton del Rio Negro who recollects Humboldt in Venezuela. He was making turtle oil on the Orinoco, on a playa near the mouth of the Apuré, when that distinguished traveller passed on his way towards the cataracts. A person died in San Fernando two or three years ago who had seen Humboldt and Bonpland at Esmeralda, and remembered the difficulty they had in procuring the flowers of the Juvia (*Bertholletia* 356 *excelsa*), for which, said he, they offered an ounce of gold. At the season of fruit of this tree the Guaharibos descend much below the raudal in order to collect it for food, and at that time the Indians of the Casiquiari, in parties of not more than five or six, lie in wait for them and carry off such as they can lay hold on, making of them slaves for cultivating their cunúcos. Many Indians on the Casiquiari can show lance-wounds received from the Guaharibos in these expeditions."

Cerro Duida

"... Duida looks down on us from the left and has seemed close by since 401 entering the Orinoco; nor has our change of position much changed its aspect till late this afternoon, when rounding a point the southern end came in view deeply cloven into four abrupt ridges. At sunset the mountain was very grand, the ridges assuming a purple hue, while the interstices were veiled in impenetrable gloom, and a stratum of white fleecy cloud was floating below the summit. The confrontation was much like that of the Serra de Curicuriarí, but less picturesque. My telescope shows that, except in a few places where the rock is very steep (whitish, sometimes streaked with brown), the mountain is forest-clad to its very summit. Yet so clear does it stand out to view, and so much nearer does it seem than it is in reality, that one would affirm its sides to be clothed with fern. Two flat summits to the north of the middle of the mountain seem to be the highest points, judging from the height of the clouds floating over them. The space below these is singularly hollowed out, and is said to be occupied by a laguna. The north extremity is a subconical peak."

"This mountain could only be climbed to the summit (if that be practicable), 431 or even to any considerable height, by sleeping two nights at the cunúco and devoting the whole of the intervening day to the ascent. But we had no provisions and there was nothing to eat in the cunúco but cassave, so that I passed a miserable night, for I had no supper and I tried in vain to sleep in a tiny hammock of very open texture, shivering with cold and tormented by zancudos, which are said to be abundant all along the base of Imei.

We started to return next morning without breakfast save a little cassave soaked in tucupí. I had torn my naked feet on the previous day and I contrived shortly after starting to deepen one wound by treading on a sharp stump, so that, what with bleeding feet and an empty stomach, I found the journey sufficiently toilsome. But this did not prevent me gathering such plants in flower as I had noted on the previous day."

Hazards of exploration

"I came worse out of this encounter than any other in which I have been 365-366 engaged since entering South America. Many times have I been stung by ants and wasps, but never so badly. Once, near São Gabriel, in my visit to the falls of Camanáos, I was making my way to a small campo; a branch hung inconveniently across my way and I made a cut at it with my cutlass, not noticing that a wasps' nest was suspended from it; but I was not left a moment longer in ignorance, for a cloud of the vile insects "buzzed out wi' angry fyke", and attacked me tooth and tail. I ran back, beating away the wasps; my hat fell off and a good many of them remained with it, but not a few still followed me, got into my hair, and stung me all over my head and neck. When I fairly got free from them I sat down on the ground, for I was dizzy and stupefied, and it seemed as if my head were bursting, for I suppose I had not fewer than twenty stings in the head and face alone. It came on to rain smartly, and I allowed the rain to beat on my head and neck, which in a few minutes seemed to relieve me much. After a while I was able to recommence my journey, though still in great pain, and I cut myself a track through the bushes so as to give a wide offing to the wasps' nest. The pain grew gradually less acute, though it did not fairly pass off all day. An Indian whom I was taking with me, and who had lagged a great way behind, had the luck to thrust his head into the same wasps' nest, and also got considerably stung ... I have been twice stung by the common house scorpion, but the pain was not greater than that produced by an English wasp. There is a larger kind whose sting is said to be far worse. The bite of the common Scolopendra (centipede) is about equal to that of the scorpion, but I have never been bitten by the immense Scolopendra that is seen in heaps of timber or among rubbish in deserted houses."

"... I calculated on spending a month in the voyage up the Casiquiari, but after 435 passing the mouth of Lake Vasiva mosquitoes began to be so abundant that my Indians became very impatient of stoppages. So long as we continued in motion, comparatively few mosquitoes congregated in the piragoa, but when we stopped to cook or gather flowers they were almost insupportable, and the cabin especially became like a beehive. You will easily understand that, however much my enthusiasm as a naturalist might conduce to render me insensible to suffering and annoyance, I could not help occasionally participating in the feelings of my sailors, and was not sorry to get along as quickly as possible. The weather was unusually fine and dry for this region; hence the abundance of mosquitoes. The same circumstance was favourable for preserving specimens, but the trees of the riverside had mostly shed their flowers and had fruit too young to be worth gathering. Still I found enough to keep me occupied."

"One of the most notable things in the Pacimoni was a tree which was con- 442 spicuous from afar by certain white cones thickly scattered along the deep green foliage. These cones my telescope revealed to be fruits, but my Indians insisted they were wasps' nests, and even when we came directly under the tree, which was not more than 40 feet high, not one of them would venture to climb it until they had first poked one of the cones with a long stick. Nor did their caution appear to me ridiculous, for on the Casiquiari we had had feeling proof that wasps' nests occur of all shapes and sizes. I expect this tree will constitute a new genus of Clusiaceae, allied to Platonia."

"Indians of the Rio Negro, Uaupés, Casiquiari, Orinoco (and perhaps of the 483-484 Amazon) eat the large grubs bred on various growing palm stems, but especially in Pihiguas. They are said to be of the size of the forefinger, and the mode of eating them is this. By a sudden twist of the head, it is pulled away along with the intestinal canal, and the animal is then roasted on the budari or mandiocca oven. There is another grub or caterpillar found on Marima trees which they are very fond of. When this insect is in season, it constitutes a principal part of the food of the Maquiritari Indians, and Don Diego Pina related to me that, travelling once

Fig. 21. A Yanomami Indian community house at Serra dos Surucucus, Roraima.

Fig. 22. Yanomami Indians together with a Brazilian field assistant, Osmarino Monteiro.

on Alto Orinoco with a crew of those Indians, he was near perishing of hunger, for they would neither fish nor seek after any sort of food but these caterpillars, and wherever they stopped by the way they climbed into the Marima trees in search of them.

I have many times seen Indians eat the saúba ant (called bacháco in Venezuela). The large kinds only are eaten, and at those times when the bachácos pour them from their holes in great numbers (probably sending forth colonies after the manner of bees), if it be near any pueblo all the unoccupied Indians in the place turn out to collect them. The head and thorax is the part eaten, the abdomen being nipped off (at San Carlos I constantly see them eaten entire), and it is eaten uncooked. The taste to me is strong, fiery, and disagreeable, but those who have eaten the bacháco fried in turtle oil tell me it is quite palatable."

"Indians are sorry nurses, and are ever more ready to flee from the sight of a sick 463 man than to help him. When they desert even their own sick relations, it can hardly be expected of them to abide by a stranger in that state. My Indians did not leave me, but I might as well have been alone. I had violent attacks of fever by night, with short respites in the middle of the day, and on the second night, on stepping out of my hammock, I was seized with vomiting, which symptom being desirous to encourage, I called to my men to heat water for me to drink. They were all so completely stupefied with rum that not one of them was able to help me. Although I had given them a bottle of rum to keep them in good-humour, I found they had sold some of my beef to obtain more. I passed a dreadful night, and in the morning I resolved to seek better aid. A friend wanted men to go to San Miguel on the Guainia, so I lent him my Indians on condition that he would find a woman who would undertake to nurse me. In the afternoon he brought with him an elderly woman who agreed to act as my nurse, but on condition of my moving to her house, where she had a family which she could not leave; and I had no choice but to agree. This woman - Carmen Reja by name - I shall not easily forget."

[Spruce was too ill to write about what followed so the editor, A R Wallace wrote:]

"During this period his nurse would often leave the house empty for six hours 465-466 at a time, evidently expecting and hoping to find him dead on her return. In the evening, after lighting his lamp and leaving a supply of water on a chair by his bed, she would often fill the house with her friends and spend the time in discussing or abusing him, calling him all the vile names in which the Spanish language is so rich. Among other things she would call out: "Die, you English dog, that we may have a merry watch-night with your dollars!" One night when the symptoms were very bad she shut up the house and did not return till long after midnight. On another evening she invited her son-in-law and other friends to spend the night with her, in the expectation (as Spruce heard her whisper to them) that the Englishman could not last out the night. Another night, when a similar termination was expected, she scolded him because he was going to leave her responsible for the safety of his goods, and one of the men whispered to her that he thought it would be necessary to give the white man some poison. At length, on the nineteenth day of the fever (July 23), a change for the better occurred, partly, he thinks, owing to his leaving off purging pills which he had taken too frequently. He now slept better, was able to eat a little, and obtained some good red wine which he took daily."

Fig. 23. A Maquiratari Indian girl.

Fig. 24. A Maquiritari Indian round house at Auaris, Roraima.

Back to Manaus

"Excursion from Barra, February 12, 1855, to the Rio Taruma. This 495-497 small river enters the Rio Negro about five hours' rowing above the city, where the coast bends inwards forming an extensive bay into which the Taruma enters. It is fairly wide at first, but as it receives numerous small streams from either side it soon becomes narrower, yet its sources are said to be a long way off in the forest. At about an hour from its mouth a rather large igarapé enters on the east side and is celebrated for having the loftiest waterfall known on the Rio Negro. My object was to visit this; and I accordingly established myself at the only Indian sitio within this branch, tenanted by an old man named Nicolas (a Manáos Indian born at Barcellos), his wife, two sons - stout lads - two grown-up daughters, and a little boy, a grandson ..."

"The next morning, accompanied by Charlie and the old Nicolas, I started to visit the fall. We ascended the winding igarapé for nearly an hour. It was much obstructed by the gapó vegetation, and at last became so grown over that we had to leave our boat and make our way through the forest. A little more than an hour brought us to the fall, which we approached from above, but we scrambled down the rocks to the bottom, where we could obtain a perfect view of the fall. I have seen few finer things in South America, and it reminded me a little of the Irish "Turk cascade". This branch of the Taruma traverses a narrow valley, contracted to a ravine below the fall, which rushes over a concave cliff in an unbroken cascade of from 30 to 40 feet high. The upper stratum of the cliff is of hard whitish sandstone, and projects considerably beyond the lower, which are of softer stone with thin alternating layers of vermilion strong-smelling earth. It is thus easy to walk under the cataract without being wetted, though the rocks drip here and there and are everywhere thickly clad with ferns and Hepaticae, but especially with Selaginellae, of which I gathered four species not found in the adjacent forests. The water falls into a deep trough, from which spray dashes out and is borne downward by the violent wind caused by the rush of the cataract. The water winds away among mossy blocks and then is lost beneath them for a considerable distance. From among these blocks springs a tree to the height of some 100 feet, the spreading sapopemas (buttresses) at its base clad by *Micropterygium leiophyllum* and a Plagiochila (Hepatics), the trunk rough with termites' nests, on which Philodendrons (Araceae) and a Carludovica (Pandaneae) have established themselves. This tree bore numerous grey fruits the size of an orange, but I could not distinguish the form of the leaves, and my guide could not give me a name for the tree, as, he said, the fruit was not edible. It was probably a Caryocar (Rhizoboleae). From top to bottom of the cataract hangs a thick rough rope of tangled black rootlets proceedings from a tree on its edge.

The whole aspect of this mossy cirque, with its broad riband of falling water, embosomed in dense luxuriant forest, in which was visible no palm, was something of an admixture of tropical scenery with that of temperate climes."

Fig. 25. Deforestation in central Amazonia, a new phenomenon since the time of Spruce.

Conclusion

[The Tarumã Falls near Manaus is a good place to leave these quotations. Richard Spruce's travels in Amazonia were indeed heroic: it is a miracle that he survived and it is incredible what a large collection of interesting plants he managed to dry and to send back to Kew. When I first visited Manaus in 1965 the Tarumã Falls were still much the same as he described them, with the carpets of ferns and *Selaginella* that he describes and even a tree of *Caryocar*. Today, 30 years later, the Falls are a small muddy trickle and the forest around them has been cleared because the Tarumã river was diverted when the new Manaus airport was built. Though many of the things described by Spruce can still be seen, many others are changing and since the early 1970s Amazonia has also been subjected to a great deal of unnecessary deforestation for unsustainable forestry and cattle farming projects. It is now unfortunately possible to see scenes which represent vast destruction. In 1993 gold miners massacred over 40 Yanomami Indians in their invasion of Yanomami Territory. I do not need to go through the details of Amazonian destruction again because they have been so well documented by the media and have been the focus of so many scientific meetings. The good news is that deforestation reached its peak in 1987 and has progressively slowed down since then. However, many problems continue, such as the integrity of the Yanomami Reservation and the land rights of the settlers and small farmers in the region.

If we really want to honour Richard Spruce, we will all work much harder for the protection of the species diversity and the cultural diversity described by him. We must not give up in Amazonia because despite all the frightening pictures of deforestation we have seen and the atrocities about which we have read, only 12 per cent of the region has been totally destroyed. There is much left for which to fight in Amazonia. Most of the biological species are still there, but Amazonia's indigenous cultures are under the severest threat and many are unlikely to survive into the 21st century. What a tragic loss of knowledge about how to manage the forests sustainably! My hope is that those who commemorated Richard Spruce did not leave the celebration feeling that the job had been done and that we duly paid homage to the great Yorkshireman, but instead returned to our home institutions more resolved than ever to help the Amazonian countries to preserve their biodiversity in every way possible, both as individuals and through pressing for more governmental financial support for the conservation and sustainable use of the Amazon region.]

9

Richard Spruce: his fascination with liverworts and its consequences

Raymond E. Stotler

Richard Spruce: his fascination with liverworts and its consequences

Raymond E. Stotler

Southern Illinois University, Carbondale, USA

Introduction

Serious students of bryophytes, especially the hepatics and anthocerotes, are quick to learn of Richard Spruce and his extensive contributions to our knowledge of this group of plants. His travels, observations, collections, and synthesis of data for nearly 15 years in the Amazon basin, its tributaries (1849–1854) and the Andes Mountains of Peru and Ecuador (1855–1864) have left us with a debt of gratitude. Humboldt and Bonpland botanized in many of these same Latin American regions from 1799–1803[1] and while their expedition has been said to be the most important ever made to America,[2] the Spruce journey is, without doubt, unparalleled in its consequences for bryology. He not only provided future generations of botanists with a wealth of information about these plants in his published works, but also left a legacy in his distribution of specimen duplicates, entitled "Hepaticae Spruceanae: Amazonicae et Andinae ... Annis 1849–1860 Lectae".[3] Indeed, access to these collections allows one to share, at least in part, some of the fascination that these plants held for Spruce, as he discovered each and every one.

The editing and publication of the manuscript notes and sketches made by Spruce during his travels to South America was undertaken by A.R. Wallace and appeared as two volumes titled *Notes of a botanist on the Amazon & Andes*.[4] Included in this work is a detailed 27-page biography which traces his career from his initial interest in nature, "... especially of the lowliest plants – the Mosses and Hepaticae – which was the joy of his early manhood and the consolation of his declining years." This is followed by a list of 54 publications attributed to Spruce. While each published account is scholarly and of notable significance, three publications in particular stand out with reference to hepatics and anthocerotes, namely, "On *Anomoclada*, a new genus of Hepaticae, and on its allied genera *Odontoschisma* and *Adelanthus*",[5] "On *Cephalozia* (a genus of Hepaticae), its subgenera and some allied genera",[6] and "Hepaticae Amazonicae et Andinae".[7] Excerpts from these works nicely illustrate the attraction that these plants held for Spruce. The following case studies are presented to demonstrate how remarkable these plants can be and, at the same time, show how Richard Spruce, with the comparatively limited facilities of his day, could make astute observations and interpretations and align seemingly unrelated taxa into natural groupings, or, conversely, cleave a large group of species within a single genus and sequester taxa into workable groups (subgenera).

Materials and methods

The following herbarium specimens, representing taxa collected and studied by Spruce, have been examined using both optical and scanning electron microscopes: *Anomoclada mucosa* Spruce (= *A. portoricensis* (Hampe & Gottsche) Vána), leg. Spruce, Silva Amazonica: S. Carlos et Javita [FH], *Arachniopsis filifolia* Spruce (nom. herb. = *A. coactilis* var. *filifolia* Spruce), leg. Spruce, Silva Amazonica, Fl. Negro et Uaupés [MANCH], *Cephalozia bicuspidata* (L.) Dum., leg. H.J.N. Schoenmakers, teste S.R. Gradstein, Obgrimbie, Belgica [ABSH], *Chaetocolea palmata* Spruce, leg. Spruce, Andes Quitenses: Tunguragua [FH], *Frullania arecae* (Spreng.) Spruce, leg. R. Düll: 3, Prov. Chiapas, Mexico [ABSH], *Frullania atrata* (Sw.) Dum., leg. Spruce, Guayrapata, Andium Quit. [MANCH], *Frullania bicornistipula* Spruce, leg. D. Griffin et al., 176, Cartago, Costa Rica [ABSH], *Frullania replicata* Nees (= *F. nodulosa* Nees), leg. Spruce, Panuré fluvii Uaupés in caatingas (= 718) [MANCH], *Frullania squamuligera* Spruce, leg. Spruce, ad Catar. Agoyán, Andes Quit., 1857 [MANCH], *Hygrolejeunea spongia* Spruce (nom. herb. = *Lepidolejeunea spongia* (Spruce) B. Thiers), leg. Spruce, Andes Quitenses: Tunguragua [MANCH], *Microfrullania duricaulis* Spruce (nom. herb. = *Frullania duricaulis* Spruce) leg. Spruce, Caripí, pr. Pará [MANCH], *Mytilopsis albifrons* Spruce, leg. Spruce, Andes Peruviana: M. Guayrapurina [MANCH], *Pteropsiella frondiformis* Spruce, leg. Spruce, Silva Amazonica, S. Gabriel et S. Carlos [MICH], *Cephalozia, Trigonanthus monodactylus* Spruce (nom. herb. = *Cephalozia* (*Zoopsis*) *monodactyla* Spruce, = *Arachniopsis monodactylus* (Spruce) R.M. Schust., = *Regredicaulis monodactylus* (Spruce) Fulf.), leg. Spruce, S. Gabriel, fl. Negro [MANCH].

Specimens prepared for SEM study were soaked in 1% TWEEN for 12–24 hrs, after which they were rinsed in dH_2O and then placed in a saturated chloral hydrate solution (250 gm chloral hydrate/100 ml dH_2O) for 5–7 days. Following two 2 hr rinses in dH_2O, the specimens were fixed in a solution of 2% glutaraldehyde and 2% formaldehyde in 0.1 M Na-Cacodylate buffer, pH of 7.2 for 12 hrs at 4°C. After rinsing with buffer they were postfixed for 3 hrs in 2% buffered osmium tetroxide. The plants were then rinsed in dH_2O and taken through a standard graded ethanol series, over a 30 hr period. Specimens were then critical point dried, using CO_2 as the transition fluid, in a Tousimis Samdri – 790 CPD. Plants were mounted on aluminium stubs using double sticky tape and coated with gold in an ISI PS2 sputter coater with a water cooled stage. Examination was with an ISI Alpha 9 counter top SEM.

Case study I – Anomoclada

Following his return to England, there was a full dozen publications by Spruce, dated 1864 and after[8] dealing with such diverse topics as cotton cultivation in northern Peru, volcanic tufa, grass fertilization, venomous reptiles and insects, and palms, to mention just a few. It was not until 1876, however, that the first paper on hepatics appeared, that being on a new genus named *Anomoclada*.[9] Reading this article, one can seemingly be transported to the Amazon basin and the Andes Mountains, so vivid is the account given; for example, Spruce[10] tells of an hepatic found on the slopes of the Andes with "... an extraordinary power of retaining the water of rains. This liverwort, "... called *Lejeunea spongia* n. sp. [placed in his new subgenus *Hygrolejeunea = Lepidolejeunea spongia* (Fig. 1)], whose densely-packed bipinnate stems form round balls, two inches in diameter, pale-green without, white within, on the twigs of trees on Mount Tunguragua, at 2500 to 3000 metres. These vegetable sponges are so constantly full of water that I have occasionally slaked my thirst by squeezing one into my mouth; for along nearly the whole northern slope of Tunguragua there is no visible stream of fresh water …". This is but one example of the vivid dialogue on hepatics that pervades all of his written works. The significance of this particular publication is that it is more than just an account of the discovery and subsequent description of several new and totally unique liverwort genera. Spruce also refined hepatic systematics in several respects. For example, here he segregated *Jungermannia carringtoni* (Balfour) Spruce and two other species into *Jamesoniella* subg. nov., and *Jungermannia donniana* Hooker and several other species into *Anastrophyllum* subg. nov. Both of these were later regarded as genera by Lees[11] and Stephani,[12] respectively. Additionally, he transferred *Jungermannia perfoliata* Sw. and several other species to *Syzygiella* gen. nov. These innovations filled several missing links among known genera, modifying the range of characters for several groups.

Of particular interest from the title, of course, was the name *Anomoclada*. Comparing his new find with the genus *Odontoschisma*, he found his plants to have larger, longer, more crisped leaves. Of greater significance were the "… leafy branches and the female flowers springing from the upper face of the stem, and not from the under, as all the branches and flowers do in *Odontoschisma* and some other allied genera …".[13] Such deviant branches obviously prompted his choice of the generic name. Perhaps of more fasincation to Spruce than the antical branching, however, was the fact that the patches he examined were always suffused with mucilage, which he thought at first to be tremelloid growth; but this was a constant feature each and every time he encountered this taxon in various regions of the Amazon basin over the next two years. He was finally convinced that the mucus was actually exuded by the plants. "It did not swell out into a jelly-like mass as a *Tremella* would have done, but looked rather as if the plant had been liberally smeared with gum arabic by means of a brush. Although I took up, with cloths, as much of the mucus as I could, the specimens still adhered so firmly to paper, especially by their underside, as not to be detached without tearing away portions of it."[14] Thus, his epithet "*mucosa*" for this anomalous plant!

Restoration of herbarium material for SEM study removes these mucus residues, and allows one to see the ventral insertion lines of the leaves and the roughened edges of the slime-secreting underleaves (Fig. 2). Evans[15] compared *Anomoclada mucosa* with *Odontoschisma portoricense* (Hampe & Gottsche) Steph., a somewhat atypical species with oblong to ligulate leaves that are crispate near the postical base, resembling poorly developed specimens of *A. mucosa*. He also pointed out that in *Odontoschisma* the underleaves possess slime papillae similar to the slime-secreting papillae on the underleaves of *A. mucosa*, regarding this, then, as a quantitative character at best and leaving antical branching as the only significant character to maintain the taxa as distinct. More recently, Schuster[16] reported sporadic *Anomoclada*-type branches in *O. denudatum*, making the distinction between the two genera even less pronounced. He did not, however, choose to combine the two genera, but rather only mentioned that the species *O. portoricense* and *A. mucosa* seem to be identical. Finally, in 1989 the new combination *Anomoclada portoricense* (Hampe & Gottsche) Vána was made, synonymizing these two species.[17]

Case study II – Cephalozia s. lat.

Perhaps because of his enjoyment of microscopy and the pleasure he found in observing the minute, Spruce spent an inordinate amount of time studying the *Cephalozia* complex. This becomes quite apparent, for example, in reading through his correspondence with Matthew Slater at Manchester Museum, in which this genus is mentioned time and time again. The allure that this genus held for him lay in the assemblage of morphological expressions that had previously been deemed of at least generic, if not of tribal, rank. In particular, he discussed such contrasting features within the genus as a frondose versus a leafy habits, succubous, transverse, and incubous leaf insertions, and acrogynous versus cladogenous fructifications. Numerous additional characters, such as branch insertion, male and female inflorescences, perianth keels, and capsule cell wall structure were addressed as well. What he perceived in *Cephalozia* was a genus embodying a remarkable series of morphological expressions, transitioning from one extreme to the other but connected, or linked together, by particular species. For example, he interpreted leaf insertions as transitioning from succubous to transverse to subincubous, or for stem leaves to disappear entirely and be replaced by a thalloid habit. These concepts supported his view of a large genus, which he partitioned into eight subgenera, five having been previously recognized as genera, the other three of which he was to name in his publication, "On Cephalozia ...".[18] These included, following his sequence, *Zoopsis* J.D. Hooker & T. Tayl., *Pteropsiella* Spruce, *Protocephalozia* Spruce, *Alobiella* Spruce, *Eucephalozia* Spruce (= *Cephalozia*), *Cephaloziella* Spruce, *Lembidium* Mitt., and *Odontoschisma* Dum. Reviewing a few of these eight subgenera nicely illustrates the above mentioned concepts held by Spruce.

First, in the generitype, *C. bicuspidata* (L.) Dum. one species sees the fundamental characters of the complex. This is not only the most common species of the genus in the British Isles, but it is likewise quite common in Europe and North America. Reviewing but a few of its features, one finds in *C. bicuspidata* (Fig. 3) bilobed leaves that are slightly succubous to almost subtransverse, underleaves absent, but with rhizoids along the postical stem surface, sexual branches postical with the female being on a very short branch, and a perianth which is trigonous with one ventral and two lateral keels. It is quite remarkable that Spruce aligned *Pteropsiella*, a plant with a thalloid habit (Fig. 4) which he likened to the fern *Pteropsis furcata*, with a genus such as *Cephalozia*. While he suggested comparison to such simple thalloid taxa as *Metzgeria* Raddi and *Pallavicinia* S.F. Gray where it might be referred, he, of course, also observed the postical sexual branches, such as the males with their broad, leafy bilobed bracts (Fig. 4), strongly resembled the perigonia and perichaetia of *Cephalozia*.

Another aberrant taxon was his new species *Cephalozia* (*Zoopsis*) *monodactyla* Spruce, described from the Rio Negro of Brazil (Fig. 6). Here, what Spruce saw was a plant with stems reminiscent of "... slender silken or silver threads ..." with minute leaves consisting "... of a single large truncato-conical basal cell, tipped by a much smaller and slenderer unguiform cell [= slime papilla] ...".[19] He, likewise, described postical branches (Fig. 7) and noted the likeness of mature vegetative plants of this species with figures 8 and 9 (Plate ix) of young *C. bicuspidata* plants illustrated by Hofmeister,[20] a similarity that is, indeed, quite striking. Still another strange plant, discovered in 1854 by Spruce, not far from the confluence of the Casiquiari and Rio Negro in Venezuela, was the monotypic *Protocephalozia ephemeroides* Spruce. At first, he took this plant to be the ephemeral moss, *Phascum* Hedw., but soon realized that he had an elaborately branching, uniseriate, persistent, *Cephalozia*-type protonema, i.e., a neotenic hepatic with persistent juvenile organization. His link, once again with *Cephalozia*, was the production from this "protonema" of leafy sexual branches that closely resembled the male and female inflorescences of *Cephalozia*, the same key that unlocked the paradox of *Pteropsiella*.

Yet one more peculiar plant discovered by Spruce in the Amazon basin at the Rio Negro and Uaupés, Brazil, and on at least one occasion found covering the roof of a cavern in the Peruvian Andes, is a genus that he felt linked *Cephalozia* with a subgenus of *Lepidozia*, namely, his subgenus *Microlepidozia* Spruce.[21] This genus consisted of very small plants "... with threadlike entangled stems, branched only from the underside, and woven into broad thin films very like a spider's web: hence my name for them, *Arachniopsis*".[22] Surprisingly, a somewhat cobwebby appearance is seen when these herbarium packets, more than 100 years old, are opened and observed with low magnification of a dissecting microscope. Three species were originally described, each having capillary leaves consisting of cylindrical cells 2–6 times as long as broad, with a succubous insertion, and with the female inflorescences being cladogenous (Fig. 5), as in *Cephalozia*. Spruce[23] mentioned the casual similarity of this genus to *Blepharostoma* (Dum. emend. Lindb.) Dum. but pointed to a close relationship only with *Lepidozia* (*Microlepidozia*) *chaetophylla* Spruce (= *Telaranea chaetophylla* (Spruce) Schiffn.).

In his publication "On *Cephalozia* ...",[24] Spruce lamented the fact that a linear arrangement of his taxa could not be effected without certain dislocation of affinities. Nevertheless, his written account nicely documented the multifarious kinship of these taxa. Soon after his treatise appeared, generic status was given to his subgenera by Schiffner,[25] who further expanded the Trigonantheae to include several newly described genera. Evans,[26] in his classic paper on the classification of the Hepaticae, placed the first of the Spruce sequence into the family Lepidoziaceae Limpr., regarding the remainder of the taxa in the Spruce list as comprising the Cephaloziaceae. This was, however, with the exception of four taxa, *Blepharostoma*, *Cephaloziella*, *Hygrobiella*, and *Pleuroclada* (= *Pleurocladula* Grolle), which he placed elsewhere. The major point of deviation between the Evans system and the most recent detailed classification scheme of Schuster[27] is the removal of still more of the Sprucean taxa from the Cephaloziaceae, hence severing their relationship with *Cephalozia*, and merging them with members of the Lepidoziaceae. Of note here are the exact genera discussed above, *Zoopsis*, *Pteropsiella*, *Protocephalozia*, and *Arachniopsis*. Schuster described a new genus of the Lepidoziaceae, *Pseudocephalozia*, in 1965[28] and

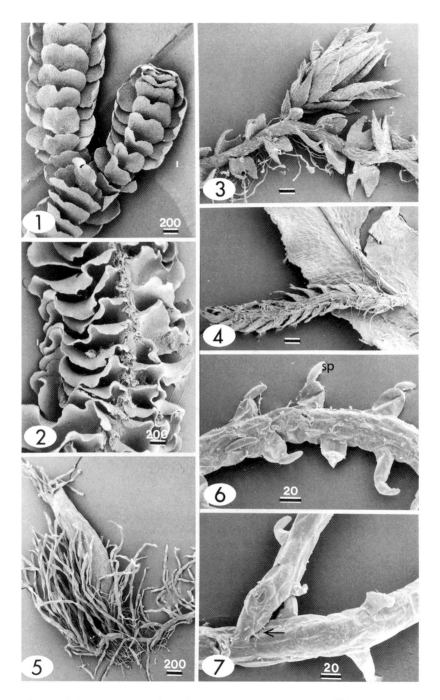

Fig. 1-7. Scale bars = 100 μm unless other μm value indicated. **Fig. 1.** *Lepidolejeunea spongia*, postical habit. **Fig. 2.** *Anomoclada portoricense*, postical habit. **Fig. 3.** *Cephalozia bicuspidata*, female plant with perianth, antical habit. **Fig. 4.** *Pteropsiella frondiformis*, male plant with postical perigonial branch, postical habit. **Fig. 5.** *Arachniopsis coactilis*, female plant with perianth, lateral habit. **Fig. 6–7.** *Regredicaulis monodactylus*. **Fig. 6.** Postical view showing single-celled leaves, capped with slime papillae (sp), with a dislodged slime cell adhering to the stem. **Fig. 7.** Postical habit showing a ventral exogenous branch at the arrow.

using this lepidozioid genus as a basis of comparison, rather than *Cephalozia*, he aligned the above taxa with the Lepidoziaceae. Refinements in classification are, however, still to come with the Sprucean taxa. A most obvious example is seen in *Zoopsis* sensu Spruce. Considering our knowledge of hepatics today, all would agree that the three species of this genus (subgenus) treated by Spruce in 1882[29] represent at least two discordant elements. The correct realignment of these elements is, however, equivocal, considering just one of these, *C. (Zoopsis) monodactyla*. Fulford[30] placed this taxon in her new genus *Regredicaulis* (*R. monodactylus* (Spruce) Fulf.) whereas Schuster[31] forced it into *Arachniopsis* (*A. monodactylus* (Spruce) Schuster) as an isolated element (cf. Figs. 5 & 6). Oddly enough, if such a reduction as that proposed by Schuster proves to be correct, hepatic systematists are likely to return to recognizing of certain of these genera as subgenera again, sensu Spruce.

Case study III – Frullania

This primarily tropical genus is second only to *Plagiochila* in richness of species names, with well over 1000 species having been attributed to it. As with *Plagiochila*, however, modern species concepts would dictate that the actual true species numbers are much, much lower with probably no more than 350–400 "real" species existing in *Frullania*. In *Synopsis Hepaticarum*, Gottsche, Lindenberg and Nees[32] treated more than 150 species world-wide, rendering the genus somewhat unmanageable. Spruce[33] carefully observed distinctions among the 47 species that he recognized from South America and established six groups based primarily upon the perianth structure, lobule size/shape and insertion, leaf lobe insertion, and the plant size and branching habit. These were given subgeneric rank[34] since "although the central and typical species of each subgenus, taken alone, might seem sufficiently distinct to justify our raising the latter to the rank of genera, the outlying species of one group merge into those of the next so that no limit is assignable ...". Today, hepaticologists still strive to solve the systematic mysteries of *Frullania*, but no better system than that proposed by Spruce more than 100 years ago has been forthcoming, although obvious refinements have been made. For example, one additional subgenus was added by Verdoorn,[35] followed by the addition of further subgenera, sections, and subsections by other students of the genus in the succeeding 65 years. To date, 18 subgenera have been named, a few of which include but a single species (e.g., *Frullania* subgenus *Orientales* Hattori, 1976[36] and *Frullania* subgenus *Huerlimannia* Hattori, 1977[37]). While not all 18 of the subgenera are currently recognized,[38] the six original subgenera described by Spruce still stand and in fact form the core of *Frullania* systematics yet today. These comprise the following: *Frullania* subgenus *Frullania* (= *Thyopsiella* Spruce), *F.* subg. *Chonanthelia* Spruce, *F.* subg. *Trachycolea* Spruce, *F.* subg. *Homotropantha* Spruce, *F.* subg. *Meteoriopsis* Spruce, and *F.* subg. *Diastaloba* Spruce.

Frullania subg. *Chonanthelia* contained 21 species, six of which were widely distributed in the plains and lower hills of the Amazon basin with the remaining 15 species being montane to subalpine in the Andes. These are usually quite large plants, with the Spruce specimens ranging from 1–4 inches to as much as 6–8 inches in length. Spruce could never find populations of these various species in the deep woods but rather they were always found associated with human habitations in the Amazonian lowlands. In the highlands, they are recorded again in association with the activities of man and domesticated animals, preferring to grow on trees at the edge of cattle and sheep pastures, or on the widely dispersed trees used for shade by cattle in the large natural pastures. Of the six subgenera named, this one is most distinct in that the perianths are 4-keeled (two lateral, two postical) with a trapezoidal configuration, sharply contrasting with the 3-keeled perianths (two lateral, one postical) in the other five subgenera. In several species, 10 to 12 supplementary folds are produced which result in modifying this geometry to produce pluriplicate perianths which still, however, have four prominent keels. Additionally, the innermost perichaetical leaves are basally connate forming what Spruce referred to as "a wide-mouthed obpyramidal vase". Vegetative characters may also serve to distinguish this subgenus, as seen in *Frullania arecae* (Spreng.) Gottsche (Fig. 8), reported by Spruce from the Andes of Ecuador. The lobules, associated with the large imbricated leaf lobes, are quite large and broad and are inflated or galeate only in the upper half, being skirted in the lower half by an explanate expanse of tissue (Fig. 9). Hence, meristematic zone position responsible for lobule inflation is uniquely different from that of all of the other subgenera. The underleaves are usually quite large (approaching the leaf lobe in size), orbicular, deeply cordate at the base, and scarcely divided (Fig. 10).

Frullania subg. *Trachycolea* mimics *Chonanthelia* species in distribution. Of the four taxa treated here, two of the species were reported from the plains and lower hills, with the other two species from the wooded Andes up to 1800 m. These are also absent from deep woods and follow the activities of man. Being quite a bit smaller, they form orbicular patches on stones and trees trunks, or are found trailing over other hepatics. The lobules are comparable to those of *Chonanthelia* in one regard: i.e., they are galeate, but here the heterogeneous growth is central in position so that the whole of the lobule becomes inflated (Fig. 12), rather than just the upper portion. It is this helmet-shaped lobule that places this subgenus nearest the *Chonanthelia*. A marked distinction is the presence of papillae, tubercles or squamules on the surface of the trigonous perianth. *Frullania squamuligera* Spruce, which was collected in the "Andes Quiteneses" at the cataract of Agoyan by Spruce (Fig. 11), nicely illustrates these subgeneric features.

Third was the subgenus *Homotropantha*, still known today in the New World by only a single species, *Frullania replicata* Nees (= *Frullania nodulosa* Nees). This nearly circumtropical species was encountered fairly frequently by Spruce throughout the Amazon basin and, according to Spruce[39] "... probably owes its wide distribution partly to the involuntary agency of man, from its almost invariably clinging to the twigs of the domesticated shrubs (such as *Crescentia Cujete*) [Calabash-tree] which accompany him in his migrations ". While the large size of the leaf lobes and the underleaves are reminiscent of the subgenus *Chonanthelia*, vegetative plants can be distinguished immediately by the presence of minute, cylindrical lobules, some, if not most, of which are reflexed ("replicate") (Figs. 13–14). In addition to these pendulous leaf lobules, the underleaves of representatives of this subgenus are extremely large, being three to six times the width of the stem as well as being broader than long.

Fig. 8-14. *Frullania* habits. Scale bars = 100 μm unless other μm value indicated. **Fig. 8–10.** *F. arecae.* **Fig. 8.** Postical habit with immature female inflorescence. **Fig. 9.** Postical habit showing lobule and stylus. **Fig. 10.** Postical habit showing mature underleaf and lateral *Frullania* type branch. **Fig. 11–12.** *F. squamuligera.* **Fig. 11.** Postical habit of female plant with mature inflorescence. **Fig. 12.** Postical habit showing underleaf and leaf lobule. **Fig. 13–14.** *F. nodulosa.* **Fig. 13.** Postical habit of female plant with mature inflorescences. **Fig. 14.** Postical habit, showing underleaves and pendent leaf lobules.

The subgenus *Meteoriopsis* reminded Spruce of mosses such as *Meteoria* [*Meteorium* (Bridel) Dozy & Molkenboer] which have a pendulous habit similar to the species he placed in this group. He writes of encounters with *Frullania atrata* (Sw.) Nees which "... depends in huge masses, sometimes half-a-yard long, and too bulky to be grasped in the arms".[40] The habit of the four species he treated is in sharp contrast to those discussed above. These species were not found at all in the Amazon basin, but rather only in the dampest hill forests of the Andes. The leaf lobes are usually convolute even in freshly collected specimens, the lobules are clavate to long cylindrical and sit parallel and close to the stem, and the underleaves are relatively narrow, being much longer than broad (Fig. 15–16).

Various species assigned to the subgenus *Thyopsiella* (= subg. *Frullania*) over the years are intermediate between *Meteoriopsis*, on the one hand, and the final subgenus of Spruce, *Diastaloba*, on the other. This, naturally, has created confusion in subgeneric delimitations which have still not been clarified in a satisfactory manner. Consequently, certain species have been shuffled among these three subgenera, or alternatively, sequestered into new subgenera. Spruce[41] gave us a working model that he based largely on his studies in the Amazon and Andes; the solution surely lies with simultaneous study of all of the species of these three subgenera worldwide. This became quite apparent in a revision of *Frullania* subgenus *Frullania*,[42] treating only Latin American taxa. Geographic and subgeneric constraints forced a very consecutive approach, resulting in the retention of intermediates in the subgenus *Frullania* where they had been previously placed by Spruce and others, even though Schuster[43] fails to give credit to these original authors. The 13 species assigned to this subgenus by Spruce[44] are absent from the Amazonian plain and are described as being wild plants that avoid the proximity of man, but are not to be found deep in the forest. In general, they are distinguished from *Meteoriopsis* by being more compact, lacking convolute leaves, having larger, broader underleaves with recurved margins, and possessing a semicordate leaf lobe base. *Frullania bicornistipula* Spruce, one of the species illustrated in his work,[45] was assigned to this subgenus (Fig. 18). While the leaf lobes and lobules are characteristic of the subgenus *Frullania*, the underleaves are not at all typical. An additional feature which has not been recognized in this species is the position of leaf lobules in relation to the underleaves. The leaf lobules develop prior to, but in close proximity to, the underleaves on one side of the stem and so are formed overlying the underleaf bases, always on the basiscopic but never on the acroscopic sides of the branches (Fig. 18). While this character can easily be seen with the low magnification of a dissecting microscope, it was overlooked until revealed with SEM micrographs. Perhaps such a character will prove to be systematically significant following the survey of additional species.

The final subgenus that Spruce described is the most problematic and the most controversial, the subgenus *Diastaloba*. The four species included by Spruce[46] agree in their small size, often distant leaves, small underleaves, short cylindrical lobules, and stylus geometry, which is triangular. These characters can be seen in *Frullania duricaulis* Spruce, collected near Pará (= Belém) near the mouth of the Amazon River (Fig. 17). Some species placed in this group such as *F. duricaulis* have lobules parallel to the stem, a character shared with the subgenus *Frullania* (cf. Fig. 17 and Fig. 18) while in others the lobules are at an obvious angle to the stem, often of about 45°. The "interlobule" (= stylus) of *F. duricaulis*, though, is triangular (Fig. 17), which explains Spruce's alignment of it with this subgenus. Schuster,[47] in contrast, would place species with a uniseriate stylus, such as *F. ecuadorensis* Steph., here because of lobule angle, while Stotler[48] retained this same species in the subgenus *Frullania* where it had been placed by Stephani. Obviously, the characters of lobule insertion angle, and stylus morphology need to be reassessed in all of the taxa that have been placed into this seemingly unnatural subgenus. For example, in my study of the type specimen (MANCH) of *F. duricaulis*, I found that the stylus is always triangular but that the lobules may be either parallel to or at an angle with the stem.

The status of *Frullania* systematics seems quite remarkable today when compared with Spruce's treatment of it. In his "conspectus generum Jubulearum",[49] Spruce aligned the genus *Jubula* with *Frullania*, considering it to be intermediate between *Frullania* and the genus *Lejeunea*. This concept had been proposed previously and has been followed by most modern authors, who recognize the Jubulaceae, comprising *Frullania* and *Jubula*, as separate from the Lejeuneaceae. Between 1962 and 1972, two species of *Frullania*, one species of *Jubula*, and two newly described taxa constituted the basis for the naming of five new genera considered intermediate links between *Frullania* and *Jubula*. However, the first of the proposed genera, *Neohattori* Kamim.,[50] based upon *F. herzogii* S. Hatt. was returned to *Frullania* by Stotler and Crandall-Stotler in 1987.[51] Next, a new genus proposed by Schuster,[52] *Jubulopsis* R.M. Schust. (based upon *Jubula novae-zelandiae* Hodgs. & S. Arn.) was shortly thereafter transferred to the remote family Lepidolaenaceae by Hamlin,[53] and *Amphijubula* R.M. Schust., named in that same paper, but based upon a new species, was synonymized by Engel[54] with *Frullania microcaulis* Gola and maintained in *Frullania*. *Steerea* S. Hatt. & Kamim.[55] based upon a new species, became synonymized with *F. clemensiana* Verd.[56] and was transferred to *Frullania* by Schuster in 1992.[57] Similarly, the final genus, *Schusterella* S. Hatt., Sharp & Mizut.,[58] based upon *F. microscopica* Pears., met with the same fate, being transferred back to the genus *Frullania* by Schuster in 1992.[59] Ironically, then, *Frullania* systematics has come full circle – from the establishment of subgenera by Spruce[60] to the recognition of five intermediate genera, back to the recognition of only the genus *Frullania*, with subgenera.

Conclusions

The impact of Richard Spruce on hepaticological study has been and still is substantial. It quickly becomes obvious that his 15 years of work in South America was more than that of a plant collector; rather it was the impassioned exploration of a scientist with an extraordinary ability to observe, record, and synthesize. While the number of his writings were relatively few, each is of lasting significance. Thiers[61] in her succinct introduction to the reprint edition of the *Hepaticae Amazonicae et Andinae* points out that within this work of nearly 600 pages only three new genera were named, whereas 39 subgenera, 374

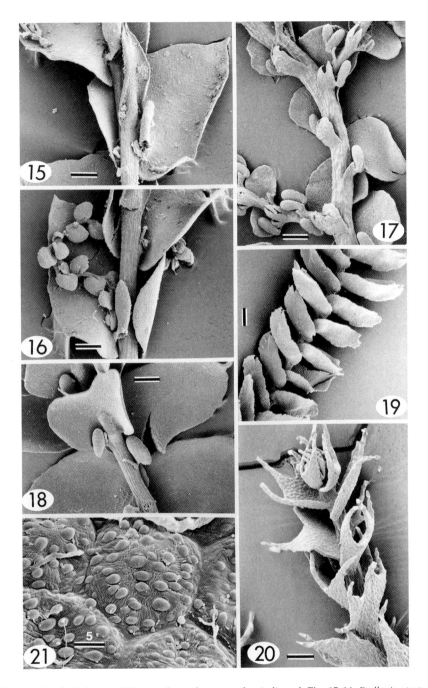

Fig. 15-21. Scale bars = 100 μm unless other μm value indicated. **Fig. 15-16.** *Frullania atrata*. **Fig. 15.** Postical habit showing leaf and underleaf. **Fig. 16.** Postical habit of stem with underleaf removed, showing a small Lejeunioid hepatic growing as an epiphyte on the ventral surface of the *Frullania* leaf. **Fig. 17.** *Frullania duricaulis*, postical habit. **Fig. 18.** *Frullania bicornistipula*, postical habit. Note leaf lobule overlying underleaf. **Fig. 19.** *Mytilopsis albifrons*. postical habit. **Fig. 20-21.** *Chaetocolea palmata*. **Fig. 20.** Antical habit. **Fig. 21.** Cell wall surfaces, showing highly ornamented wall.

species and 132 varieties were described as new to science. Not only here, but in his other publications as seen above, the rank of genus novus was seldom afforded. Spruce practised the ultimate in conservative systematics, with most of his subgenera today being regarded as *bonafide* genera.

Although his South American work contained detailed descriptions of more than 700 species and varieties, a total of only 24 taxa were illustrated on 22 full-page plates inserted at the end of the work. Spruce did not prepare these drawings himself; plates 1–8 were made by the British bryologist Robert Braithwaite, while plates 9–22 were drawn by George Massee, a British bryologist at Kew who later travelled to South America at Spruce's suggestion.[62] Plate 14, *Mytilopsis* Spruce, represents what Spruce called a curious and beautiful plant which was in every way like the genus *Micropterygium* save for the lack of under-leaves. The generic name for this plant came from his observation of "... leaves so equally and closely complicate that they resemble in miniature a slightly-gaping bivalve shell, such as that of the mussel ...".[63] (Fig. 19). Finally, one of the many extraordinary hepatics Spruce was to find was *Chaetocolea palmata,* figured on plate 12.[64] This monotypic genus was placed by Fulford[65] in its own family, the Chaetocoleaceae, which is regarded today as a monotypic subfamily of the Pseudolepicoleaceae Fulf. & J. Tayl. by Grolle[66] or of the Herbertaceae K. Müll. by Schuster.[67] Vegetative plants are characterized as having leaves quadrifid to the middle (sometimes trifid) with similar, although somewhat smaller under-leaves (Fig. 20). Unlike many mosses, liverwort cell walls are typically smooth whereas the cell walls of this hepatic were, as illustrated by Massee and described by Spruce[68] "... minute verruculoso-striolata". Richard Spruce, the microscopist, would no doubt have marvelled were he able to view this unusual ornamentation with the scanning electron microscope (Fig. 21).

Epilogue

Realizing how very few botanists today study hepatics, let alone hepatics in South America, it is a great tribute to Richard Spruce that his *Hepaticae of the Amazon and the Andes of Peru and Ecuador* was reprinted in 1984 – approximately 100 years after it first appeared.[69] Although written primarily in Latin, this work still remains the standard reference for this geographic region. Adulation was paid when Schuster[70] published an appreciation of Spruce. This included a fitting quote from a Spruce letter to Daniel Hanbury in 1873 which, according to Wallace,[71] was "In reply, apparently, to some deprecatory remarks upon his favourite Hepatics ...". The letter reads, in part: "I like to look on plants as sentient beings, which live and enjoy their lives which beautify the earth during life, and after death may adorn my herbarium. When they are beaten to pulp or powder in the apothecary's mortar they lose most of their interest for me. It is true that the Hepaticae have hardly as yet yielded any substance to man capable of stupefying him, or of forcing his stomach to empty its contents, nor are they good for food; but if man cannot torture them to his uses or abuses, they are infinitely useful where God has placed them, as I hope to live to show; and they are, at the least, useful to, and beautiful in, themselves – surely the primary motive for every individual existence."

As a lasting tribute to Richard Spruce, two hepatic genera were named in his honour, both of which stand today. These are *Sprucella* Stephani (1886) and *Spruceanthus* Verdoorn (1934).

ACKNOWLEDGEMENTS

First and foremost, I should like to thank my wife, Barbara Crandall-Stotler, for her mastery in rendering herbarium specimens, some more than 100 years old, suitable for SEM preparation and for her expertise with the microscope. Additionally, I extend my gratitude to Sean Edwards for the loan of critical Spruce collections from Manchester University Museum and for his kind hospitality during our visit; to Jerry Snider of the Cincinnati University Herbarium for literature and specimens; Donald Pfister of the Farlow Herbarium, Harvard and Howard Crum, University of Michigan for herbarium collections, and finally Robert Kiger, Hunt Institute for Botanical Documentation, for providing me with valuable bibliographic data.

NOTES

(1)   D. Botting, *Humboldt and the cosmos*, New York, 1973.

(2)   W.T. Stearn, 'Humboldt and Bonpland's "Voyage aux régions équinoxiales"', in *Humboldt, Bonpland, Kunth and tropical American botany* (ed. W.T. Stearn), Stuttgart, 1968, 3–7.

(3)   G. Sayre, 'Cryptogamae exsiccatae. An annotated bibliography of exsiccatae of algae, lichenes, hepaticae and musci. V. Unpublished exsiccatae. 1. Collectors', *Memoirs of the New York Botanical Garden* (1975), **19**, 401–402.

(4)   R. Spruce, *Notes of a botanist on the Amazon and Andes* (ed. A.R. Wallace), London, 1908, 2 vols.

(5)   R. Spruce, 'On *Anomoclada* a new genus of Hepaticae, and on its allied genera, *Odontoschisma* and *Adelanthus*', *Journal of Botany*, (1876), **14**, 129–136, 161–170, 193–203, 230–235 + pl. clxxvii-clxxix.

(6)   R. Spruce, *On Cephalozia (a genus of Hepaticae), its sub-genera and some allied genera*, Malton, 1882.

(7)   R. Spruce, 'Hepaticae Amazonicae et Andinae', *Transactions and Proceedings of the Botanical Society of Edinburgh*, (1884-1885), **15**, 1-588. [pp. 1-308 published April, 1884; pp. 309-588 published November, 1885. Reprinted as 'Hepaticae of the Amazon and of the Andes of Peru and Ecuador', 1885; reprinted 1984 as *Contributions from the New York Botanical Garden*, **15**, with an introduction and index with updated nomenclature by B.M. Thiers.]

(8)   Spruce, op. cit. (4)

(9)   Spruce, op. cit. (5)

(10) Ibid., 131–132.

(11) F.A. Lees, *The London catalogue of British mosses and hepatics*, 2nd edn, London, 1881.

(12) F. Stephani, 'Hepaticarum species novae, II', *Hedwigia* (1893), **32**, 137–147.

(13) Spruce, op. cit. (5), 129.

(14) Ibid., 129.

(15) A.W. Evans, '*Odontoschisma macounii* and its North American allies', *Botanical Gazette* (1903), **36**, 321–348 + pl. xviii-xx.

(16) R.M. Schuster, *The Hepaticae and Anthocerotae of North America, east of the hundredth meridian*, New York, 1974, vol. 3.

(17) S.R. Gradstein, 'A key to the Hepaticae and Anthocerotae of Puerto Rico and the Virgin Islands', *Bryologist*, (1989), **92**, 329–348.

(18) Spruce, op. cit. (6)

(19) Ibid., 11–12.

(20) W.F.B. Hofmeister, *On the germination, development and fructification of the higher Cryptogamia and on ... the Coniferae* (transl. F. Currey), London, 1862.

(21) Spruce, op. cit. (5).

(22) Spruce, op. cit. (6), 16.

(23) Spruce, op. cit. (6).

(24) Ibid.

(25) V. Schiffner, 'Hepaticae (Lebermoose),' in *Die natürlichen Pflanzenfamilien* (eds. A. Engler and K. Prantl), Leipzig, 1893, 1 Teil, 3 Abteilung, 1–141.

(26) A.W. Evans, 'The classification of the Hepaticae', *Botanical Review* (1939), **5**, 49–96.

(27) R.M. Schuster, 'Evolution, phylogeny and classification of the Hepaticae', in *New Manual of Bryology* (ed. R. M. Schuster), Nichinan, l984, vol. 2, 892–1070.

(28) R.M. Schuster, 'Studies on Hepaticae. XXVI. The Bonneria-Paracromastigum-Pseudocephalozia-Hyalolepidozia-Zoopsis-Pteropsiella complex and its allies: a phylogenetic study (part 1)', *Nova Hedwigia* (1965), **10**, 19–61.

(29) Spruce, op. cit. (6)

(30) M.H. Fulford, 'Manual of the leafy Hepaticae of Latin America. Part III', *Memoirs of the New York Botanical Garden* (1968), **11**, 277–392.

(31) Schuster, op. cit. (28).

(32) C.M. Gottsche, J.B.W. Lindenberg and C.G.D. Nees von Esenbeck, *Synopsis Hepaticarum*, Hamburg, 1844–1847. [Pp. 1–144, 1844; pp. 145–464, 1845; pp. 465–624, 1846; pp. 625–834, 1847.]

(33) Spruce, op. cit. (7)

(34) Ibid., 6–7.

(35) F. Verdoorn, 'V. Schiffner, expositio plantarum in itinere suo indico annis 1893/94 suscepto collectarum speciminibusque exsiccatis distributarum, adjectis descriptionibus novarum. De Frullaniaceis IV', *Annales Bryologici* (1929), **2**, 117–154.

(36) S. Hattori, 'Notes on the Asiatic species of the genus Frullania, Hepaticae. X', *Journal of the Hattori Botanical Laboratory* (1976), **40**, 461–507.

(37) S. Hattori, 'Dr. H. Hürliman's collection of New Caledonian Frullaniaceae', *Journal of the Hattori Botanical Laboratory* (1977), **43**, 409–438.

(38) R.M. Schuster, *The Hepaticae and Anthocerotae of North America, east of the hundredth meridian*, Chicago, 1992. vol. 5.

(39) Spruce, op. cit. (7), 36.

(40) Ibid., 38.

(41) Spruce, op. cit. (7).

(42) R.E. Stotler, 'The genus *Frullania* subgenus *Frullania* in Latin America', *Nova Hedwigia* (1969), **18**, 397–555.

(43) Schuster, op. cit. (38).

(44) Spruce op. cit. (7).

(45) Ibid.

(46) Ibid.

(47) Schuster, op. cit. (38).

(48) Stotler, op. cit. (42).

(49) Spruce, op.cit. (7), 2.

(50) M. Kamimura, 'A monograph of Japanese Frullaniaceae', *Journal of the Hattori Botanical Laboratory* (1961), **24**, 1–109.

(51) R.E. Stotler and B. Crandall-Stotler, 'A re-evaluation of the genus *Neohattoria* (Jubulaceae)', *Memoirs of the New York Botanical Garden* (1987), **45**, 535–543.

(52) R.M. Schuster, 'Studies on antipodal Hepaticae, III. *Jubulopsis* Schuster, *Neohattoria* Kamimura and *Amphijubula* Schuster', *Journal of the Hattori Botanical Laboratory* (1970), **33,** 266–304.

(53) B.G. Hamlin, '*Jubulopsis* Schust., with notes on athecal branching in leafy Hepaticae', *Journal of the Hattori Botanical Laboratory* (1973), **37**, 169–180.

(54) J.J. Engel, 'A taxonomic and phytogeographic study of Brunswick Peninsula (Strait of Magellan) Hepaticae and Anthocerotae', *Fieldiana, Botany* (1978), **41**, viii + 1–319.

(55) S. Hattori and M. Kamimura, 'A new genus of Frullaniaceae (Hepaticae) from Borneo', *Journal of the Hattori Botanical Laboratory* (1971), **34**, 429–436.

(56) S. Hattori, 'On the status of little-known North Bornean *Frullania, F. clemensiana* Verd. (Hepaticae)', *Miscellanea Bryologica et Lichenologica* (1975), **7**, 13–14.

(57) Schuster, op. cit. (38).

(58) S. Hattori, A.J. Sharp and M. Mizutani, '*Schusterella*, a new genus of Jubulaceae (Hepaticae)', *Japanese Journal of Botany* (1972), **20**, 329–338.

(59) Schuster, op. cit. (38)

(60) Spruce, op. cit. (7)

(61) B. Thiers, 'Introduction', in, Spruce, op. cit. (7) [1984 reprint], ix–xii.

(62) cf. F.A. Stafleu and R.S. Cowan, *Taxonomic Literature*, 2nd edn, Utrecht, 1981, vol. III: Lh–O.

(63) Spruce, op. cit. (6), 90.

(64)   Spruce, op. cit. (7)

(65)   M. Fulford 'Manual of the leafy Hepaticae of Latin America. Part 1', *Memoirs of the New York Botanical Garden* (1963), **11**, 1–172.

(66)   R. Grolle 'Nomina generica Hepaticarum; references, types and synonymies', *Acta Botanica Fennica* (1983), **121**, 1–62.

(67)   Schuster, op. cit. (27)

(68)   Spruce, op. cit. (7)

(69)   Spruce, op. cit. (7) [1984 reprint].

(70)   R.M. Schuster, 'Richard Spruce (1817–1893): a biographical sketch and appreciation', *Nova Hedwigia* (1982), **36**, 199–208.

(71)   A.R. Wallace, in, Spruce, op. cit. (4), xxxviii.

10

Spruce's *Hepaticae Amazonicae et Andinae* and South American floristics

S. Rob Gradstein

Spruce's *Hepaticae Amazonicae et Andinae* and South American floristics

S. Rob Gradstein

University of Utrecht Herbarium,
Heidelberglaan 2, Utrecht, The Netherlands

Hepaticae Amazonicae et Andinae[1], a 600-page monograph on the hepatics of the Amazon basin and the Andes, is Richard Spruce's *magnum opus*. The book appeared in 1884-85, about twenty years after his return from South America. Today, in spite of being over a hundred years old, *Hepaticae Amazonicae et Andinae* is still the most important reference on tropical South American hepatics. Spruce wrote his monograph in the village of Coneysthorpe, Yorkshire, where he lived after his return to England until his death in 1893. The work was published in two instalments in the *Transactions & Proceedings of the Botanical Society of Edinburgh* and was distributed in book form in 1885 by Trübner & Co., London, under the title *Hepaticae of the Amazon and of the Andes of Peru and Ecuador*.[2] A reprint of the latter edition was published in 1984 by the New York Botanical Garden.

Hepaticae Amazonicae et Andinae was not the first work on neotropical hepatics. About a century earlier the first hepatic species had been described from this part of the world by the Swedish physician and botanist Olov Swartz. In the course of the 19th century, extensive papers had been published on the liverworts of Mexico and Colombia by C.M. Gottsche, of Brazil by G. C. Nees von Esenbeck, and of French Guiana and Bolivia by C. Montagne. A monograph on the important genus *Plagiochila* had been written by J.B.G. Lindenberg. By the time Spruce left for South America, almost three hundred species had been described from the neotropics.[3] These, however, were based on the study of herbarium material and none of the authors, with the exception of Swartz, had seen the tropics, nor had they been able to observe the species in the field, study their growth forms and branching patterns, their colours, habitats, etc. As Thiers[4] wrote: "*Hepaticae Amazonicae et Andinae* displays a level of sophistication far beyond that of Spruce's predecessors or contemporaries in the field of 'exotic hepaticology'". The fact that Spruce knew all his species in the field, together with his outstanding skills as a taxonomist, his eye for detail and his passion for the hepatics, must be reasons why his book is so good.

The taxonomic merits of the work are discussed elsewhere.[5] Many of Spruce's taxonomic concepts are still valid today and modern classifications of important families such as the Frullaniaceae, Plagiochilaceae and Lejeuneaceae are heavily based on Spruce's treatment in *Hepaticae Amazonicae et Andinae*. Although some of his species have fallen into synonymy, the majority of his new taxa are still accepted. Some of them have not been collected since their discovery by Spruce and three species, the curious *Myriocolea irrorata* Spruce (Lejeuneaceae), *Spruceanthus theobromae* (Spruce) Gradst. (Lejeuneaceae) and the moss *Fissidens hydropogon* Spruce ex Mitt. (Fissidentaceae), all from Ecuador, have been placed on the world list of threatened taxa.[6]

In this chapter, *Hepaticae Amazonicae et Andinae* is reviewed in the light of our present-day floristic knowledge of the area. It should be noted, however, that for most of the areas visited by Spruce his work remains the only floristic treatment available, the notable exception being a study of the bryophytes of the surroundings of Tarapoto.[7] Consequently, the subject cannot be discussed in a very elaborate manner.

The Amazon basin is often considered one of the richest areas in the world in terms of numbers of species. However, for hepatics and some other plant groups, this does not seem to be true. In a letter published recently in *Nature*, Henderson *et al.*[8] have pointed out that for many plant groups the centre of diversity is in the Andes, not in Amazonia. In the case of ferns, there seem to be at least five times as many species in the Andes than in the Amazon basin. With respect to hepatics, the authors contend that the Andes have about four times more species. The latter figure is somewhat questionable, however, as it was based on a checklist of Colombia which is rather incomplete for lowland areas.[9] It would seem more appropriate to evaluate hepatic richness in the two areas based on *Hepaticae Amazonicae et Andinae*.

In his travel logs and letters published posthumously,[10] Spruce makes frequent mention of the very different hepatic floras of the Amazon basin and the Andes. In the hot Amazonian lowlands he found mostly species of the family Lejeuneaceae, *Plagiochila* and *Bazzania* species, one species of *Lophocolea* and a few other minor groups. In the Andes, Lejeuneaceae appeared to become scarcer with increasing elevation and were replaced by a variety of other groups. In general, hepatic diversity at the family and generic level appeared much greater in the mountains than in the lowlands and this has also been observed by later workers. It seems that in the tropics hepatics are generally better adapted to growth in cooler mountain climates, where hazards of desiccation are less and habitats more suitable. Furthermore, immigration of temperate groups from higher latitudes should also have added to the greater taxonomic diversity of hepatics in the mountains.[11]

With respect to species richness, Spruce recorded about 560 species (including almost 400 new ones) from Amazonia and the Andes. This is a very good tally because even today the total number of "good" species in the region, excluding possible synonyms and other superfluous names, is probably less than a thousand. While a few of the species collected by Spruce occurred both in the Andes and in Amazonia, the great majority were restricted to only one of these areas. Of the 560 species found by Spruce, 381 or about 68% occurred in the Andes, whereas 213 or about 38% occurred in the Amazon basin. Thus, it would appear that the Andes had about 1.8 times as many species than the lowlands. This ratio is considerably lower than the one of Henderson *et al.*[12] and is also lower than the 2-3:1 ratio found by Gradstein.[13]

The discrepancy between *Hepaticae Amazonicae et Andinae* and the recent publications mentioned above is worth a closer examination. Why did Spruce find so many species in Amazonia, and so few in the Andes? He spent several years collecting in each of the regions and there is no evidence that he paid less attention to hepatics while in the Andes than during his stay in the Amazon basin. Although the area travelled in Amazonia is a much larger one in terms of surface area than the area explored in the Andes, the smaller area in the Andes is compensated for by the far greater elevational variation.

I believe that two factors are particularly relevant here. Firstly Spruce paid much attention to the rich flora of the upper Rio Negro and Orinoco. With respect to hepatics, this region yielded twice as many species as the other parts of Amazonia he visited. Thus, it would appear that the number of species recorded from Amazonia would have been much smaller had he not visited this area. Secondly, it should be realised that his collections from the Andes are almost exclusively from below the timber line. Although Spruce did collect in the alpine belt or *páramo*, he found this zone to be quite poor in hepatics: "La zone alpine des Andes est aussi pauvre en hépatiques que celle des Pyrenées et l'espèce qu'il a vue monter le plus haut, c'est le *Jung. cordifolia* Hook., qui est aussez commune dans les montagnes de l'Europe".[14] The *Jungermannia*, which was gathered on the Pinchincha near Quito at 4000 m, is indeed the only species in *Hepaticae Amazonicae et Andinae* recorded from páramo vegetation; it has been determined as *J. ovato-trigona* (Steph.) Grolle, a South American páramo endemic, by Vána.[15] The lack of páramo records in *Hepaticae Amazonicae et Andinae* is surprising because more than 250 species, including numerous endemics, have been recorded from the páramos of the northern Andes.[16] Genera such as *Anastrophyllum, Diplophyllum, Gymnomitrium, Isotachis, Jensenia, Lophozia, Marsupella, Pseudocephalozia, Ruizanthus, Stephaniella* and *Triandrophyllum*, are hardly or not at all treated in the book, yet they are quite common in the páramos and some (*Anastrophyllum, Marsupella, Gymnomitrium*) are represented by a rather large number of species. Why Spruce did not collect these is not entirely clear but it may be due to the fact that the páramo areas which he visited (Pichincha, Chimborazo) were relatively dry and therefore poor in hepatics. Had he been able to collect in the humid páramos of the northern Andes, he would undoubtedly have been able to gather many more species and, as a result, the number of Andean species treated in *Hepaticae Amazonicae et Andinae* might have been considerably higher.

In spite of this floristic gap, *Hepaticae Amazonicae et Andinae* remains the most comprehensive treatment of the hepatics of the Amazon basin and the Andes and for many areas visited by Spruce it is the only floristic treatment available to date.[17] Notable exceptions are the surroundings of Tarapoto, on the eastern slopes of the Andes in northern Peru, where Spruce collected during 1855-57. This area was re-inventoried in September-October 1982 by a team of European bryologists, the "BRYOTROP" project.[18] Although the localities inventoried by BRYOTROP were not usually the same as those visited by Spruce, the area seems sufficiently small and floristically homogeneous to allow for a comparison. As pointed out by Gradstein,[19] the Andean hepatic flora is fairly uniform and many of the local species may be found in a small area, or even only on a few trees. Elevation, however, plays an important role determining the distribution of the species.[20] Since Spruce did not collect above 6000 feet in the Peruvian Andes, BRYOTROP data from above this elevation are not taken into consideration.

A comparison between the species recorded from the surroundings of Tarapoto by Spruce and by BRYOTROP reveals striking similarities as well as differences

Table I. Hepaticae and Anthocerotae recorded from the Peruvian Andes around Tarapoto (ca. 1000-6000 feet) by Richard Spruce and by BRYOTROP; ( ) = recorded above 6000 ft., + = unidentified species.

	SPRUCE	BRYOTROP
ADELANTHACEAE		
Adelanthus Mitt.	1	( )
ANEURACEAE		
Riccardia S.F. Gray	2	5
ANTHOCERATACEAE		
Anthoceros L.	1	1
Megaceros Campb.	1	-
BALANTIOPSIDACEAE		
Isotachis (Lehm. & Lindenb.) Gott.	-	1
Neesioscyphus Grolle	1	1
CALYPOGEIACEAE		
Calypogeia Raddi	2	4
CEPHALOZIACEAE		
Alobiella (Spruce) Schiffn.	2	-
Alobiellopsis Schust.	1	-
Cephalozia (Dum.) Dum.	3	1
Odontoschisma Dum.	1	( )
CEPHALOZIELLACEAE		
Cephaloziopsis Schust.	2	-
DENDROCEROTACEAE		
Dendroceros Nees	1	1
GEOCALYCACEAE		
Heteroscyphus Schiffn.	2	1
Leptoscyphus Mitt.	2	( )
Lophoclea (Dum.) Dum.	5	7
HAPLOMITRIACEAE		
Haplomitrium Nees	1	1
HERBERTACEAE		
Herbertus S.F. Gray	1	( )
JUBULACEAE		
Frullania Raddi	11	+
Jubula Dum.	-	1

	SPRUCE	BRYOTROP
JUNGERMANNIACEAE		
Anastrophyllum (Spruce) Steph.	1	( )
Cryptochila Schust.	-	1
Jungermannia L.	1	1
LEJEUNEACEAE		
Anoplolejeuna (Spruce) Schiffn.	-	1
Aphanolejeunea Evans	-	1
Archilejeunea (Spruce) Schiffn.	3	3
Bryopteris Nees	2	2
Ceratolejeunea (Spruce) Schiffn.	3	5+
Cheilolejeunea (Spruce) Schiffn.	3	2
Cololejeunea (Spruce) Schiffn.	1	3
Colura (Dum.) Dum.	-	2
Cyclolejeunea Evans	2	3
Cyrtolejeunea Evans	-	1
Dicranolejeunea (Spruce) Schiffn.	-	1
Diplasiolejeunea (Spruce) Schiffn.	1	2
Drepanolejeunea (Spruce) Schiffn.	2	1
Echinocolea Schust.	1	1
Frullanoides Raddi	1	1
Harpalejeunea (Spruce) Schiffn.	3	3
Lejeunea Lib.	8	8+
Lepidolejeunea Schust.	1	3
Leptolejeunea (Spruce) Schiffn.	1	-
Lopholejeunea (Spruce) Schiffn.	1	2
Macrolejeunea (Spruce) Schiffn.	1	1
Marchesinia S.F. Gray	1	1
Mastigolejeunea (Spruce) Schiffn.	2	1
Microlejeunea Steph.	1	2
Neurolejeunea (Spruce) Schiffn.	-	1
Odontolejeunea (Spruce) Schiffn.	1	1
Omphalanthus Nees	1	1
Prionolejeunea (Spruce) Schiffn.	3	4
Rectolejeunea Evans	-	1
Stictolejeunea (Spruce) Schiffn.	1	2
Symbiezidium Trevis.	1	1
Taxilejeunea (Spruce) Schiffn.	4	2+
Trachylejeunea (Spruce) Schiffn.	1	1
LEPIDOZIACEAE		
Arachniopsis Spruce	2	1
Bazzania S.F. Gray	10	+
Kurzia von Martens	1	( )
Lepidozia (Dum.) Dum.	2	( )
Micropterygium Lindenb. *et al.*	4	3
Mytilopsis Spruce	1	-
Telaranea Spruce ex Schiffn.	1	1
Zoopsidella Schust.	1	1

	SPRUCE	BRYOTROP
MARCHANTIACEAE *Marchantia* L.	2	2
METZGERIACEAE *Metzgeria* Raddi	3	8?
MONOCLEACEAE *Monoclea* Hook.	1	1
NOTOTHYLADACEAE *Notothylas* Sull.	1	-
PALLAVICINIACEAE *Symphyogyna* Nees & Mont.	3	3
PLAGIOCHILACEAE *Plagiochila* (Dum.) Dum.	19	18
PORELLACEAE *Porella* L.	2	2
RADULACEAE *Radula* Dum.	4	10
TRICHOCOLEACEAE *Trichocolea* Dum.	1	2
WIESNERELLACEAE *Dumortiera* Nees	1	1

(Table 1). The total number of taxa recorded is surprisingly similar: 151 species in 64 genera by Spruce, 160 species (including several unidentified ones) in 68 genera by BRYOTROP. Most of the genera recorded in the two treatments are the same. Five genera were only found by Spruce (*Alobiella, Alobiellopsis, Cephaloziopsis, Mytilopsis, Notothylas*) as compared with nine recorded exclusively by BRYOTROP (*Aphanolejeunea, Colura, Cryptochila, Cyrtolejeunea, Dicranolejeunea, Isotachis, Jubula, Neurolejeunea, Rectolejeunea*). *Jubula* and *Neurolejeunea* are rare hepatic genera in the Andes whilst *Cryptochila, Isotachis* and *Dicranolejeunea* are high altitude taxa not commonly occurring below 6000 feet. This might explain their absence in the Spruce gatherings from Tarapoto. The other genera recorded by BRYOTROP but not by Spruce are common elements of Andean lower montane forests. The fact that Spruce did not collect these seems purely accidental. Genera found exclusively by Spruce and not by BRYOTROP are mostly uncommon taxa of specialised habitats such as steep cliffs. They are unlikely to have become extinct in the Tarapoto area but may be locally rare.

As to the actual species recorded, the two inventories show strikingly different results. Of the 151 species found by Spruce, only about 42 or less than one third were also recorded by the BRYOTROP. Only five species of *Plagiochila*, out of more than thirty reported in the two treatments, are the same, only one *Lejeunea* (sixteen reported) and none of the *Ceratolejeunea* species (nine reported)! On the other hand, of twelve species of Geocalycaceae recorded in total, eight were reported both by Spruce and by BRYOTROP and of ten species of ptychanthoid Lejeuneaceae recorded by Spruce nine were also gathered by BRYOTROP. The only ptychathoid missing was *Archilejeuna porelloides* (Spruce) Schiffn., a species which Spruce recorded as rare around Tarapoto. The Geocalycaceae and Ptychanthoideae are taxonomically quite well known and their nomenclature is fairly straightforward.[21] The other groups are still in a state of nomenclatural chaos, however. It seems obvious, therefore, that the differences at species level are to a considerable extent due to insufficient taxonomic knowledge. Other reasons may have been alteration of habitats since the days of Spruce and the different localities visited by Spruce and by BRYOTROP. The latter fact may have been the reason why local rarities such as *Archilejeunea porelloides* (Spruce) Schiffn. were not collected by BRYOTROP.

To what extent habitat alteration has influenced the results remains speculative. Although the mountains around Tarapoto are still covered by forest, some deforestation and expansion of farm land is apparent in the area and this is likely to have affected the local hepatic flora. In particular, "shade epiphytes", characteristic of the understorey of the forest, may have become rarer.[22] The available data seem to be inadequate to allow for conclusions on this point, however. A more detailed investigation of the local hepatic flora, including study of the actual localities visited by Spruce and analysis of habitat requirements of individual species, would have been necessary to elucidate the possible impacts of the environmental changes.

The conclusion that I would like to draw from this brief review is that unless we make more progress with taxonomic revisions of neotropical hepatics and analyse the possible impacts of habitat alteration, it remains very difficult to evaluate the diversity of the hepatic flora of tropical America and the changes that might have occurred since the days of Spruce. Because of the speed at which the natural habitats are being destroyed in this part of the world, such basic taxonomic studies are more urgently needed today than they have ever been since Richard Spruce did his pioneering work on the hepatics of the Amazon basin and the Andes.

ACKNOWLEDGEMENTS

I am grateful to Prof. J.-P. Frahm for helpful comments on the draft manuscript.

NOTES

(1)   R. Spruce, 'Hepaticae Amazonicae et Andinae', *Transactions and Proceedings of the Botanical Society of Edinburgh* (1884-1885), **15**, i-xi, 1-588. [pp.1-308 published in 1884, pp.309-588 in 1885.]

(2)   R. Spruce, *Hepaticae of the Amazon and of the Andes of Peru and Ecuador*, London, 1885. [Reprint of 1884-85 edition, op. cit. (1). Reprinted 1984 as *Contributions from the New York Botanical Garden*, **15**, with an introduction and index with updated nomenclature by B.M.Thiers.]

(3)   C.M. Gottsche, J.B.G. Lindenberg and C.G. Nees von Esenbeck, *Synopsis Hepaticarum*, Hamburg, 1844-1847.

(4)   B.M. Thiers, 'Introduction', in 'Hepaticae of the Amazon and the Andes of Peru and Ecuador'(R. Spruce), [reprint], *Contributions from the New York Botanical Garden* (1984),**15**, IX-XII.

(5)   R.E. Stotler, 'Richard Spruce: his fascination with liverworts and its consequences', in *Richard Spruce (1817-1893), botanist and explorer*, (eds. M.R.D. Seaward and S.M.D. FitzGerald), Kew, 1996, 123-140.

(6)   S.R. Gradstein, 'Threatened bryophytes of the neotropical rain forest: a status report', *Tropical Bryology* (1992), **6**, 83-93; P. Geissler, B. Tan and T. Hällingback, 'Towards a World Red List of Bryophytes', Poster, XV International Botanical Congress, Yokohama, 1993.

(7)   W. Frey (ed.), 'Moosflora und -vegetation in Regenwäldern NO-Perus. Ergebnisse der Bryotrop-Expedition nach Peru 1982', *Beihefte zur Nova Hedwigia* (1987), **88**, 1-159 .

(8)   A. Henderson, S.P. Churchill and J.L. Luteyn, 'Neotropical plant diversity', *Nature* (1991), **351**, 21-22.

(9)   S.R. Gradstein and W.H.A. Hekking, 'Studies on Colombian cryptogams IV. A catalogue of the Hepaticae of Colombia', *Journal of the Hattori Botanical Laboratory* (1979), **45**, 93-144.

(10)   R. Spruce, *Notes of a botanist on the Amazon and Andes* (ed. A.R. Wallace), London, 1908, 2 vols.

(11)   R.M. Schuster, 'Phytogeography of bryophytes', in *New manual of bryology* (ed. R.M. Schuster), Nichinan, 1983, vol. 1, 463-626.

(12)   Henderson et al., op. cit. (8).

(13)   S.R. Gradstein, 'Hepatic diversity in tropical Andean Forests', *Memoirs of the New York Botanical Garden*, in press.

(14)   R. Spruce, 'Voyage de Richard Spruce dans l'Amerique equatoriale pendant les années 1849-1864', *Revue Bryologique* (1886), **13**, 77. [Paper reprinted under the alternative title: 'Précis d'un voyage d'exploration botanique dans l'Amerique equatoriale, pour servir d'introduction provisoire à son ouvrage sur les hepatiques de l'Amazone et des Andes, par Richard Spruce', p.1-20.]

(15)   J.Vána, 'Studien über die Jungermannioideae (Hepaticae). 2. *Jungermannia* subg. *Jungermannia*', *Folia Geobotanica et Phytotaxonomica* (1973), **8**, 255-309.

(16)   S.R. Gradstein, unpublished checklist.

(17)   For example, R.C. Lisboa, 'Histórico da Briologia na Amazonia Brasileira', *Boletim do Museu Paraense Emilio Goeldi, Bot.* (1991), **7**, 69-77.

(18) Frey, op.cit. (7).

(19) Gradstein, op. cit. (13).

(20) J.H.W. Wolf, *Ecology of epiphytes and epiphyte communities in tropical montane rain forests of Colombia,* Dissertation, University of Amsterdam, 1993.

(21) M.H. Fulford, 'Manual of the leafy Hepaticae of Latin America. Part 4', *Memoirs of the New York Botanical Garden* (1976), **11**, 393-535; S.R. Gradstein, 'The Ptychanthoideae of Latin America: an overview (Studies on Lejeuneaceae sub-family Ptychanthoideae, XVI)', *The Bryologist* (1987), **90**, 337-343.

(22) Gradstein, op. cit. (6); S.R. Gradstein, 'The vanishing tropical rain forest as an environment for bryophytes and lichens', in *Bryophytes and lichens in a changing environment* (eds. J.W. Bates and A.M. Farmer), Oxford, 1992, 234-258.

11

Two letters from Spruce to Braithwaite about the illustrations to *Hepaticae Amazonicae et Andinae*

P.W. Richards

Two letters from Spruce to Braithwaite about the illustrations to *Hepaticae Amazonicae et Andinae*

P. W. Richards

Wootton Way, Cambridge, UK

Two letters from Richard Spruce, one dated December 9, 1883 and the other August 18, 1884, are of interest for the light they throw on the preparation of the illustrations for *Hepaticae Amazonicae et Andinae*, as well as on Spruce's life and health – he was 67 in 1884. The 'messenger' mentioned at the end of the first letter was probably the young girl who assisted his housekeeper at Coneysthorpe. Both letters are written in pencil: most of his writing at this time was done in a reclining chair with a large book on his knees.

In the book, the illustrations are all at the end of part 2. Only the first eight (Tab. I – VIII) are from Braithwaite's drawings; the rest (Tab. IX – XXII) are from drawings by the mycologist, George Massee, who was also a friend of Spruce. Though Spruce was himself a capable draghtsman and had drawn the illustrations for his paper on the mosses and hepatics of the Pyrenees (1850), he may have felt that after the stroke he had suffered in the Andes in 1860 he was no longer able to make accurate drawings, even though his handwriting was still beautifully clear and legible.

Why did Massee replace Braithwaite after the latter had drawn the first eight species? Was Braithwaite for some reason unable to continue or was Spruce (as the letter of August 1866 perhaps suggests) not satisfied with his work? Massee's drawings are better in some respects than Braithwaite's. It is also perhaps relevant to note that the second letter was written after the first part of the book was published. Braithwaite may have thought he could not complete the remaining drawings in time for the publication of part 2.

Braithwaite's drawings for Tab. I and II were lithographed by F. Huth (referred to in the 1883 letter) and the others are inscribed 'E. Carter sc.' and 'Mintern Bros. imp.'.

Coneysthorpe, Malton
9 Dec. 1883

Dear Dr Braithwaite

You have laid me under great obligation by your excellent figures of my hepaticae, which will really make very pretty pictures. If I had been near you I might have aided you in the selection of the objects. Although no pressure was applied to the dissections, the perianths, which were quite entire when I cut them off, must have split at the apex by repeated moistening and drying; and the dehiscence of that organ is of far less importance than the number and nature of the folds, or keels, before rupture. Thus the perianth of <u>Fr. bicornistipula</u> is trigonous throughout its length, (like all the rest of its fellows in the same subgenus) while that of <u>Fr. sphaerocephala</u> is 10–12-plicate only above the middle.

You have correctly noted the interlobular process – called a <u>stylus</u> IN 'Syn.Hep.' It is found throughout the genus, usually along with the leaves on one side of the stem only and even then not with any constancy.

Along with this letter I will try to send you the first 80 pages proof-sheets of my work, but you must please return them by the end of the week, as I have often to refer to them & I have no other copy. You shall of course have a copy of the entire work when completed. On pp 4 & 42 you will see something about the stylus. On p 17 is the descr[n] of <u>Fr. sphaerocephala</u> & on p. 46 that of <u>Fr. bicornistipula</u>. In trimming & paring down the latter for publication, I see that I have cut out, not only all mention of the stylus, but also the desc[n] of the form of the <u>lobule</u>. This again indeed varies so little throughout the subgenus (<u>Thyopsiella</u>) being always subcylindrical – nunc conico-cylindrica, nunc clavato-cylindrica – that it only merits special description in rare cases.

I think you may be interested to read my remarks on the subgenera of <u>Frullania</u>.

My Edinburgh friends wish me to have the lithography and printing done in Edinburgh by Mr. Frederic Huth, who does both. I have had several spec[ns] of his work sent me, both of <u>engraving</u> on the stone (like yours) & of chalk-work & certainly in both kinds it is as good as need be. He had done a good deal for the Challenger Exped[n,] all the plates for Maw's forthcoming monograph of Crocus, etc., I therefore confided to him these two first plates, but if they do not turn out up to my mark, then I must have recourse to your engraver.

I have more to say but must write again, for I am too late for our village post, and I must send up to Castle Howard, and I do not wish my messenger to be out after nightfall.

With kind regards and renewed thanks, believe me

Faithfully yours
Rich[d] Spruce

Did you ever gather mosses in this neighbourhood?

Coneysthorpe
18 Aug. 1884

Dear Dr Braithwaite

Herewith I send specimens for figuring on Tabs. VII and VIII. The plants are rather larger than those of the two previous plates; still I fear you may find them very delicate and fragile.

Tab. VII, <u>Trachylejeunea asperiflora</u> Hep.A.A.183, is monoicous. The ♂ inflorescence is of short julaceous branches, with 5–12 pairs of turgid equilobed leaves (bracts), each (in an early stage) enclosing 2 antheridia that speedily fall away. Perianths clavate or pyriform, 5-angled, strongly papillose, especially on the angles. They are mostly burst at the apex, & the capsule fallen out, but if you look at one in a drop of water, without pressure, the sutures will close again, and a figure ought to be made of a closed and another of a split perianth. A section would be a desirable addition, but that I cannot well send.

Tab. VIII, <u>Ceratolejeunea microrhegma</u> l.c.p. 209. The peculiarity of this subgenus is that the angles of the perianth are prolonged upwards into horns. Very often only 4 of the 5 angles are horned, the antical and flatter angle being hornless. I have very little of this plant, & I regret to see that [♡ section] [*Spruce has sketched a 5-sided figure and put it in the margin in brackets*] I have only a single perfect perianth (the one enclosed) the few others having the horns more or less mutilated. The specific name is drawn from the minute cleft at the apex of the underleaves, most other Ceratolejeuneae having them cloven to about the middle. The lower leaves on a branch have often the 2 basal leaves almost entirely inrolled into a sac (utriculus). See descr^n of subgenus and introduction to Lejeunea p. 66.

My main occupation – & a very hard one too – during this long spell of sultry weather, has been to barely keep alive. I have got out of doors as much as possible and that has been limited chiefly to sitting in my garden. I have further delayed asking you to resume your work on my plants, because I hoped to see you ere this in Yorkshire, as you proposed, when, although you c^d not have drawn any of my plants: away from your own "tools", etc., I c^d have pointed out to you certain of their features liable to be overlooked by any one not familiar with hepaticae.

W^d you care to try your hand at making some additions and corrections to your figs in Tabs. V and VI? If so I will send them, along with specimens. There were fine ♂ plants of <u>Lej. ancistrodes</u> (t.V) which seem to have escaped your notice. The ♂ spikes are branchlets, clad with turgid leaves, and should have been shown in a magn^d fig. of the whole or part of a ♂ plant. I have been looking over a good many figures of hepaticae & I came to the conclusion that it is best to represent them as <u>opaque</u> objects – at least to show in the drawing only the surfaces next to the observer. If the opposite surface requires to be shown, it should be in a separate figure (see Hooker's Musci Exot. etc.)

What is Davies' flippant note worth? If the <u>Plagiochila</u> PROVISIONALLY named <u>Pl. Braithwaitei</u> be really the same as <u>Pl. Seemanniana</u> (which is very possible) Mitten ought to have been able to say so positively, for I presume the latter species is of his inventing. My note after examining the Aneiteum plant was "Pl. Bantamensi aff.", but as I was sending you Sande-L.'s Hep Javan. where there is no fig. of Pl. Bantamensis but of the not-very-remote Pl. Sandei there is a figure. I referred you to the latter for an example of an allied large-leaved species. You know I dare say what became of Seemann's herb[m]. If it is in the British Museum perhaps you c[d] get me a sight of Pl. Seemanniana. At any rate you c[d] get from the Linn. Soc. library the loan of that part of the Flora Vitiensis which contains a description of the plant.

As to the other plant, which Mitten calls Bryopteris fruticosa, if you refer to p. 110 of my work you will see that I have shewn that species to be no Bryopteris at all, & have proposed it as a new subgenus, Dendrolejeunea – not however without some doubt as to whether it sh[d] be merged in Thysananthus, & I believe it quite distinct from fruticosus by the simply pinnate (not bipinnate) stems, the branches never forked beneath the ♀ flowers, the evittate leaves with a far smaller & differently-shaped lobule etc, etc. I call it pro tem. Dendrol[a] (or Thysananthus) abietina, but I have only very young perianths, & if you have only very young perianths, and if you have any more of the plant I sh[d] like to see the whole of it – perhaps I might find more advanced perianths – & I w[d] return the spec[ns] [specimens]. I enclose one of the 2 stems I possess to enable you to identify the species – please return it.

Faithfully yours
Rich[d] Spruce

12

Cinchona work in Ecuador by Richard Spruce, and by United States botanists in the 1940s

William B. Drew

Cinchona work in Ecuador by Richard Spruce, and by United States botanists in the 1940s

William B. Drew

Professor Emeritus, Department of Botany and Plant Pathology
Michigan State University, E. Lansing, Michigan, USA

After eight years of arduous and successful collecting of plants in the Amazon basin and a portion of the Orinoco system, Spruce thought of extending his operations westward towards the eastern slopes of the Peruvian and Ecuadorean Andes.[1] He had amassed a magnificent collection of several thousand species with special attention being paid to his favourite group, the Hepaticae, but he was also particularly attracted to the ferns as well as the palms, which are among the most difficult plants, other than cacti, to collect and convert to herbarium specimens. Spruce seems to have possessed an uncanny ability to find rare species since his collections abound with them. For instance, he is credited with discovering many new species of the rubber-yielding genus *Hevea* at a time when the production of latex from wild trees was accelerating in pace with the rapid increase in the uses of the product.

Late in 1857, while collecting hundreds of plants in the vicinity of Tarapoto, Peru, Spruce received notice from Her Majesty's Secretary of State for India of a commission to proceed to the *Cinchona* forests of Ecuador,[2] there to study the species producing quinine and to obtain the best seeds and plants for shipment to India. At that time all the quinine on the world market came from the bark of wild species of *Cinchona* growing principally in Colombia, Ecuador, Peru and Bolivia. Stringent laws and severe penalties awaited anyone caught while attempting to smuggle seeds or plants out of Bolivia and Peru. By 1857, not one of many foreign agents had succeeded in doing so.[3] Yet, in Ecuador, Spruce was able, after much correspondence and diplomatic haggling, literally to rent the *Cinchona* forests of Alausí and Limon on the western slopes of the Andes, where it was generally believed by Ecuadorians that the finest and most productive species grew. Such an unrestricted permit to collect seeds and plants of *Cinchona* would have been totally impossible to obtain in Peru or Bolivia, so that Spruce's successful diplomacy with the somewhat reluctant Ecuadorian owners of the *Cinchona* forests was in itself an outstanding coup. The way was clear for him to collect seeds of the "roja-roja", or red-barked *Cinchona*, and to grow plants from some of them.

To reach Ecuador from Tarapoto, Peru, 500 miles of dangerous, difficult terrain had to be traversed, much of it by river using dugout canoes.[4] Spruce encountered heavy rains and swollen rivers as he made his way from the Rio Hullaga (Peru) to the Rio Pastaza (Ecuador). It was often necessary to build bridges to cross streams in flood, but it was not always possible to find the large bamboos which would be easier to use for construction. Moreover, as he neared the lower Andean slopes, steeper gradients prevailed so that the streams became fast slowing torrents in which anyone unfortunate enough to fall would have been swept away to certain death. At times Spruce and his Indian helpers had to build one bridge after another as rising waters washed away their construction. At one such crossing on the Rio Topo, Spruce could not build a bridge strong enough to permit the passage of his heavy boxes which had to be abandoned along with a quantity of his irreplaceable collecting papers. Adding to these difficulties was a scarcity of game and edible plant products so that Spruce and his men suffered severely from hunger.

During this arduous trip, Spruce was nevertheless able to collect bryophytes to add to his growing collection. Once he reached Baños, he was rejuvenated by the abundance of food available. His spirits were further lifted by the opportunity to collect for several months the Hepaticae on and in the vicinity of the Volcán de Tungurahua, an inactive volcanic cone with a tiny girdle of ice and snow at its very summit at an elevation of slightly more than 5,000 m.

Despite the hardships Spruce endured in traversing the 500 difficult and dangerous miles from Tarapoto, Peru to Baños, Ecuador, his health was much better than it had been in Amazonia. Yet on April 10, 1860, while in Ambato, he suddenly became deaf in his left ear; but that was a minor inconvenience compared to the devastating paralysis of his back and legs which struck him 19 days later. He was never able thereafter to sit up straight, or walk without great pain. Nevertheless, he was able to initiate his *Cinchona* studies in June, seeking to determine the best types of bark, and analysing the ecological conditions relating to each type. No one had previously learned as much about *Cinchona* and its growth requirements, so that these data were immensely important to the establishment of the *Cinchona* plantations in India. During all his painful, debilitating illness at this time, he was able to get around on horseback or with a long staff to continue collecting bryophytes and ferns, as well as studying the kinds of *Cinchona* and their bark characteristics.

Spruce found that the best fruiting capsules of "roja-roja" were those from the vicinity of Limon. He had started for the *Cinchona* forests on June 12, 1860, reaching the city of Guaranda on the 14th, and Limon on the 18th. Here he found fruiting sprouts of "roja-roja" from trees harvested years before 1860, but there were very few trees left unharvested in the surrounding forest: today there is no longer a forest left on these slopes. His centre of operations was a sugar cane mill with an upper storey dormitory; there is still such a structure in existence, where in 1944 I spent an uncomfortable night on a bare board bed a foot too short for me! However, this mill is doubtless of more recent origin than the structure used by Spruce many years earlier. The principal cane product marketed in Guaranda and Riobamba was and is alcohol ("puro"), since the transport by mule or horse of crude bulk sugar over the treacherous mountain trail would not be as remunerative. Indeed, Spruce described in vivid terms the treacherousness of the mountain trail to Guaranda, marked by dead animals which had succumbed to the rigours of the exhausting experience of carrying heavy loads through deep mud and often successfully skirting precipitous slopes.[5]

Spruce paid men to guard the maturing capsules which were ready for harvest between August 13 and 18. After harvest, he opened the capsules, each of which usually contained 40 seeds which he carefully dried. He calculated that his harvest from Limon and vicinity amounted to about 100,000 ripe *Cinchona* seeds. During this period, Spruce was joined by a Mr. Cross, a British gardener, sent out by the authorities to help with the production of live plants from cuttings and from seed.[6] The net result was the successful production of 637 living plants which were carefully packed in Wardian cases for the long voyage to India. The plants arrived at their destination in good condition ready for transplanting. These plants, plus those from germinated seed, were used to establish the *Cinchona* plantations which provided the quinine to alleviate the suffering of thousands of malarial victims in the British Colonies. Now, at last, people living and working in these areas could look forward to a dependable drug to keep the dreaded malaria under control, thanks to a modest, dedicated Yorkshireman.

His great labours on *Cinchona* completed in Ecuador, Spruce embarked from Guayaquil in May, 1864, nearly 15 years after he had landed in Belém. Through his tireless efforts in Andean Ecuador, the *Cinchona* plantations of India and Ceylon, made possible by his work, saved thousands of lives over many years. In the years following Spruce's return to England in 1864, the Dutch government managed to obtain seeds of a species of Bolivian *Cinchona* (*C. ledgeriana*) which produced a relatively large amount of quinine (up to 13%) in its bark. This species of *Cinchona* became the foundation for the plantations on Java, but ironically, to produce the best yield, the *C. ledgeriana* shoots had to be grafted upon rootstocks of "roja-roja"! The Javanese plantation production of quinine led to the gradual demise of the wildbark industry of Andean South America, and to a Dutch world monopoly which lasted until the outbreak of World War II.

Thus it was that when Japan invaded Malaya and Java, about 90% of the world's supply of quinine became unavailable to the allies, whose forces were fighting in malaria-infested areas from the Mediterranean to Burma, the Solomons and the Philippines. In the meantime, drug companies in the U.S. were striving to formulate an effective synthetic anti-malarial drug based upon an original German discovery. Atabrine was found to be effective, but there were side-effects. Consequently, the U.S. Office of Economic Warfare (later the Foreign Economic Administration) developed a programme in Andean South America to attempt to revive the moribund wild *Cinchona* bark industry. Botanists, of whom I was one, and foresters were recruited to conduct exploration for stands of *Cinchona* trees, to study the different species and, with assistance from analytical laboratories, determine the quinine content of each one.

Besides *Cinchona officinalis* varieties, *C. pitayensis* of southern Colombia and northern Ecuador proved to be particularly valuable, having a quinine content of 3-4%. Substantial quantities of *Cinchona* bark were harvested and shipped to the U.S. for extraction of quinine for the allied forces.

Finally, seedlings and saplings of high-yielding strains of *Cinchona ledgeriana* were grown from seed brought out from the Philippines on the last American plane (a B-17 Flying Fortress) to leave the islands before the Japanese completed their conquest. Several Latin American countries, including Ecuador, were given these plants so that they might develop their own plantations to produce quinine. Very little effective action was unfortunately ever taken by any government.

Now, of course, there are several synthetic anti-malarial drugs on the world's markets; but quinine itself is still needed to treat specific strains of malaria.

Whenever the history of quinine is told, the name of Richard Spruce will be in the forefront for his very significant pioneering work and the careful studies which led to the first successful plantations of *Cinchona* and relief to countless thousands of people infected or threatened with infection with malaria in India and around the world.

NOTES

(1) R. Spruce, *Notes of a botanist on the Amazon and Andes* (ed. A.R. Wallace), London, 1908, vol. 2.

(2) R. Spruce, *Report on the expedition to procure seeds and plants of the Cinchona succirubra, or Red Bark Tree*, London, 1861, 3.

(3) M.B. Kreig, *Green medicine: the search for plants that heal*, Chicago, 1964, 141-192.

(4) R.E. Schultes, 'Richard Spruce still lives' *Northern Gardener* (1953), **7**, 87-89.

(5) Spruce, op. cit. (2), 67.

(6) Ibid., 9.

ACKNOWLEDGMENTS

The writer is greatly indebted to the Department of Botany and Plant Pathology, Michigan State University for a grant to help defray travel costs and to Prof. Mark Seaward for his courtesies and assistance during the conference.

13

Spruce's great
contribution to health

Plutarco Naranjo

Spruce's great contribution to health

Plutarco Naranjo

European Academy of Medicine, Quito, Ecuador

Malaria is an old disease. It did not spread as rapidly as small-pox, measles or the bubonic plague but it slowly and progressively advanced from the African tropics to other regions of the world. It claimed millions of human lives. Even the great conquerors were not immune to the disease. According to some historians, Alexander the Great met death by the microscopic parasite of malaria, following the expansion of his dominion as far as Babylonia.

Probably the first person to study the consequences of malarial infection, although not aware of it, was the famous Greek physician, Hippocrates, who classified fevers as tertian and quartan if the interval between the fever attacks was two or three days apart respectively. Many centuries later it was discovered that these fevers were caused by *Plasmodium falciparum* and *P. vivax* respectively.

When the epidemic reached Rome, it was given the name *malaria* (meaning bad air) because it was thought that the corrupt air coming from the swamps surrounding part of the city was the cause of the disease.

The parasite also reached the New World, transported by the African slaves. According to some reports,[1] it first appeared in the Caribbean Islands, but within only a short period of time, in 1586, it had reached the territory of Peru.

Discovery of quine

The early history of the use of quine to cure malaria is complex; the traditional story (questioned by some researchers) is as follows.

The treatment of malaria in the Quito Royal Audience was the same as in Spain: blood-letting, purgatives, infusions and bizarre diets. Many were saved from the disease but not all from the treatment. In Loja, in the southern part of Ecuador, in spite of the official treatment, a Jesuit was dying of a high fever, which at that time was called tertian. The Jesuit's Indian servant insisted on bringing one of Malacato's medicine men named Pedro Leiva.[2] The healer did not know anything about 'malaria', but he had a significant amount of experience in treating fevers. He gave the patient a very bitter powder, which he had diluted with chicha (the Indian corn beer). On the third day the Jesuit recovered; he was completely cured.

Not long afterwards, news came to Loja from Peru that the Viceroy's wife, the Countess of Chinchon, was sick with the horrible disease of malaria. The local governor, Juan Lopez de Canizares, diligently obtained some quina bark and immediately sent it to Lima.[3] It became apparent that the sick person was not the Viceroy's wife but the Viceroy himself, and the malady was not malaria, but bloody stools. Nevertheless, the quina was used as it was in use for various fevers in hospitalized patients. Father Calacha reported later that the powder from the "fever tree" was doing 'miracles' in Lima.[4]

The Spanish Crown, duly informed of these facts, ordered a shipment of a large quantity of the drug for the Royal Pharmacy of Madrid. The Jesuits, on their own initiative, sent a trunk full of quina powder to their superior in Rome, Cardinal Lugo. He became the "saviour" of many poor patients in that city and as a result the drug became popularly known as the Cardinal's powder.

By this time some of the New World's gold mines had already been exhausted. Quina bark thus became a new treasure. The Spanish Crown claimed for itself the monopoly of the bark. Thus began the "quina fever". Many Spaniards and "criollos" respectfully acknowledged the King's order, but as frequently occurs, many did not obey him and started an active smuggling trade.[5]

Forests were irrationally exploited and no effort was made to reafforest. Additionally, either through ignorance or deliberate fraud, if the harvest of the quina bark was poor, barks of other trees like quinquine were shipped instead, with a consequent lack of improvement in the patients and discredit of the drug. The need of a serious study to identify the real botanical species and varieties of the quina tree emerged, not only in Spain but in other countries as well.

In 1746, on the authorization of the King of Spain, the French Geodesic Mission arrived in Quito.[6] Its main purpose in coming was to measure a segment of the meridian, north and south of the equator. This was to solve an important problem, not only from the scientific point of view but also from the practical aspects of marine navigation, to determine the real shape of the earth. This group hoped to end the angry controversy between the English scientists who backed Newton's theory that the earth was flat at the poles and the French scientists, who backed Cassini's theory that the planet was elongated at the poles.

The French Mission, however, had some undeclared goals, including an investigation of the quina trees, the botanical species, and the environment of the forests as well as the gathering of seeds and other materials. A botanist, Joseph Jussieu, accompanied the Mission and was the first scientist to study the quina forests in southern Ecuador. Some time later, La Condamine, a member of the French group, also visited the quina area and collected seeds that were later lost during the journey.

Perhaps stimulated by this expedition or jealous of its success, the Spanish Crown organized a new expedition in 1776, with botanists Hipólito Ruíz and José Pavón.[7]

As the consumption of quine increased, Spain indulged in more extensive exploration to locate new quina forests. In 1783, the Royal Botanical Expedition, headed by José Celestino Mutis, was organized to travel through Nueva Granada (today's Colombia). The Expedition succeeded in discovering new forests, including some new species of Chinchonas. Unfortunately, these trees were not of high quality compared to the Loja quine.

Then came competition from the Germans. In 1799, Humboldt organized his expedition, accompanied by the botanist Aimé Bonpland, who likewise travelled to the quina country to carry out important observations and make collections.[8]

If great interest was generated by the botanical study of quine, it became even greater with the extraction in 1820 of the principal chemical, by two French chemists, Joseph Pelletie and Joseph Caventou. The alkaloid that was isolated was named quinine. This discovery opened a new avenue for medicine, that of etiological treatment or the use of a pure chemical compound for specific treatment of a disease.

The extraction of quinine, which required large amounts of raw materials, came at a period when there was a shortage of the bark. By 1810 the quine trade was facing difficulties, not only because of the deforestation but also because of the independence movements that had begun in South America and would last for more than twenty years.

Decades before, in 1760 in Quito, a famous physician and writer, Eugénio de Santacruz y Espejo, advocated the rational exploitation of quina and demanded immediate reafforestation. However, the colonial authorities ignored his plea. When Humboldt observed the exploitation of the forest, he wrote:

> "If the governments in South America do not attend to the conservation of quine, either by prohibiting the felling of trees or by obligating the territorial magistrates to force the loggers to protect them from destruction, this highly esteemed product of the New World will be swept from the country". [9]

Spruce and the quine

As frequently happens with new and "miraculous" drugs, quine was not only used to treat malaria but also for all kinds of fevers and other ailments, becoming supposedly a panacea.[10] The plant gained the name "tree of life", and the demand for the drug continued to increase at a time when the forests were almost depleted. The Spanish government needed large amouns of quine to treat the workers in the swampy areas of Lorca, Riofrio and elsewhere, and for the miners of Almad an Teruel. Since Spain had declared war against France, it especially sought quine for the army located in Cataluna, Rosellon and Navarra. The government purchased 9,5000 pounds of quine at the high price of 23 "reales de vellon" per pound.

The price of quine in England and in the rest of Europe had reached unprecedented levels, but of greater concern was the shortage of the drug. Britain, whose colonial ambitions were seriously impeded by the prevalence of malaria, was desperate to establish her own reliable source of quine. Since the beginning of the "quine crisis" as early as 1790, the shortage of the drug was causing severe problems, not only in Europe and South America but all over the world.

Richard Spruce, a tall, lean Englishman of gentle manners but characterized by a great perseverance and a steely temper, had, for more than a decade, already explored and studied the botany and anthropology of more than 3000 miles of the Amazon basin. No other European scientist had demonstrated such long term endurance of inclement weather, daily attacks by myriads of insects, continuous risk of being killed by Indians, or the tropical diseases (including malaria), which put him at the edge of death.

Spruce was still enjoying the beauty of the jungle and its enchantments for a botanist while confronting its perils, when he unexpectedly received a royal command to acquire and export to India plants and sufficient seeds of red quine. Spruce summarized the letter as follows:

"Toward the end of the year 1859, I was entrusted by Her Majesty's Secretary of State for India with a commission to procure seeds and plants of the Red Bark tree, and I proceeded to take the necessary steps for entering on its performance".[11]

No doubt it was an honour for him to be entrusted by Her Majesty, but what a commission!

Probably the people in London believed that for anyone already in the upper Amazon, to extend the journey to the quine country was relatively simple – but that was far from reality. The easiest way was probably to return to the east coast via the Amazon and then make a round trip from the Caribbean across the Isthmus of Panama to the Pacific Ocean, but this would have required far more time to prepare than was available.

When he received his commission, Spruce was in Tarapoto in the Peruvian Amazon, about 1000 miles from the quine country. To sail down the Amazon was like driving a comfortable car on a paved highway in comparison to crossing the dense jungle and mountain by mule. To reach the quine forest, it was necessary first to pass through dangerous territory, including that of the Jivaro tribe, the fearful headhunters of the Ecuadorian eastern jungle and then finally to climb the Andes mountain ranges. Many fellow-countrymen would probably have declined such an honour, while others, not having the experience of living in the Amazon area, would have laughed at the idea – but Spruce did not hesitate. He knew that his country required his service and he was keenly aware that millions of lives were depending on him. So, just as in the first days of his Amazon river expedition, he began to organize a fresh conquest of another unknown world.

On the quine country

About three hundred years earlier, the Spanish conqueror, Gonzalo Pizarro, with a group of two hundred Spaniards and about three thousand Indians, departed from Quito to the "cinnamon country" in the Amazon basin. Almost two years later and after countless troubles, Pizarro returned with only 110 Spaniards and no Indians. The jungle had devoured them. Now it was Spruce's turn; one man in contrast to two hundred and five to seven Indians in contrast to three thousand.

Part of the journey included canoeing down the Bobonaza, Pastaza and other rivers; part included walking through the jungle, sometimes using shoes and sometimes going barefooted to cross the swamps. He was travelling in one of the most humid areas of the world. In fact it rained almost all the time. Spruce reported:

"June 15th: we had heavy thunder-showers from 2 to 4 A.M., and wet dripped from the roof on to the foot of my bed ... June 16th: Again heavy showers before daylight which left the forest soaking wet for our journey ... June 17th: At daybreak rain again came on and continued without intermission till near noon ... [The Púyu river was] so swollen that there was no hope of crossing it ...".[12]

However, the delay had some compensation for a botanist. Spruce, ever the keen muscologist, wrote: *"My chagrin at this delay was somewhat lessened by the circumstance of finding myself in the most mossy place I had yet seen anywhere"*.

Despite occasional scientific compensation, according to his own words, Spruce suffered more during this arduous trip than at any time during the previous ten years. Finally he reached Quito, but he still had not reached the Chinchona country. New and different problems had to be faced: first, he had to travel through Ecuador and identify the places where the quine forests were located; second, an even more difficult and time-consuming challenge was to find a farm where he could work and wait until the trees produced ripe seeds; third, he had to attempt to cultivate the trees; and finally, most complicated of all, to arrange for exportation of the valuable material.

Part of Spruce's success was due to a physician, James Taylor, a native of Cumberland, who had lived in Ecuador for almost 30 years. He happened to be the medical attendant to General Flores, during his presidency in Ecuador. Dr Taylor was also a lecturer in anatomy in the University of Quito. Because of the local conditions and the distance involved, it was impractical to attempt to collect the material in the southern mountains of Loja. The most logical procedure was to gather the Chinchona seeds on the western slopes of the Andes. The areas chosen belonged to General Flores and the Church, the latter's land being rented to a Dr Neyra. It took several weeks of bothersome negotiation to make a treaty with them. Finally with a payment of 400 dollars, Spruce was allowed to take the seeds and the plants but not any bark.

For months Spruce explored the new forests. He made many trips but frequently only found destroyed Chinchona trees. When he did find two regions with the appropriate trees, it was not the right season for harvesting the seeds. Almost two years were required for him to gather ripe seeds, dry them, plant them and then wait for the plants to grow. At the same time he tried to reproduce the plants by cuttings. It worked. Happily the Englishman watched the cuttings sprout leaves, but soon the caterpillars, the insects and mould began to compete with him.

During the last and most critical period, although the civil war started, and Spruce had to work amongst soldiers and weapons, he was effectively assisted not only by Dr Taylor but also by Mr Robert Cross, a Kew gardener. At last more than 600 plants and 100,000 seeds were ready to be shipped by river to the main Ecuadorian port, Guayaquil, an enterprise not easily accomplished. For transportation, the group had to build a special raft. After overcoming all kinds of obstacles, Spruce arrived at the port with the materials in good condition. After a few days, on the 2nd of January 1861, the treasure was shipped to Kew and to India. Independently of this cargo, an additional large number of seeds was also sent to Jamaica.

The young plants and seeds reached India without any major problem. The seeds germinated and were used to start extensive plantations on the Nilgherri Hills in South India, in Sri Lanka, in Darjeeling and other places. Unfortunately, after seven years, it was realized that the locations chosen were unsuitable for Chinchona trees and the production of the red bark was very low. Nevertheless, this exercise stimulated the Dutch government to plant the seeds in Indonesia, where they succeeded in growing productive trees.

As a corollary to this feat Spruce wrote:

"... a far greater source of anxiety to me were the contretemps ... that every now and then threatened to bring our work to naught. It is difficult for those who live in a country of peace and plenty, but above all of good roads, to appreciate the obstacles that beset all undertakings in countries where none of those blessings exist".[13]

The long days of heavy rains, the long years of loneliness, were however propitious for meditation, expressed by Spruce as follows:

"I have often wished I could get some consular appointment here, were it only of £150 a year, but I have no powerful friends, without which a familiarity with the country, the inhabitants and the languages, go for little. A person is much wanted to watch over the interests of Europeans on the Upper Amazon, but I can hardly suggest a station for him which could not be liable to some objection, and an itinerating consul is something I have never heard of, though it would really be very useful here. The Brazilians have a vice-consul in Moyobamba. The French have a vice-consul in Santarem and another in the Barra do Rio Negro".[14]

After Spruce's expedition and the development of seedlings and cultivation of cuttings, the gross production of the bark and the industrial production of its alkaloid began. England and the Netherlands were now able to supply the drug to the whole world, with the expectation that the terrible disease of malaria would be eradicated forever. Perhaps only penicillin can be compared with quinine in regard to the number of lives it has saved. It accounts for millions.

However, the *Plasmodium* parasites have begun to develop their own resistance to the drug. Today neither quinine, synthetic choroquine nor any known antimalarial products have succeeded in eradicating the disease. According to the latest UNICEF reports,[15] 2,000,000,000 people live in malarious areas and every year more than 100,000,000 new cases result, with more than one million deaths.

Today, 150 years after Spruce's expedition, it is still not easy to cross the jungle and the Andes mountain range. No one has attempted to repeat his feat. Only a sense of loyalty to his country and above all to humanity, could have motivated Spruce to undertake this incredible adventure.

The story is told of another man, named Robert Talbor (or Talbot), who lived in London almost two centuries earlier. A garrulous employee of a pharmacy, he claimed to be the only man to be able to cure malaria, a very common disease in England at that time. In fact, he successfully treated many patients and became very rich and famous. In 1678 he was summoned to treat King Charles II. Subsequently he was knighted and appointed as Physician to the King. Later, after Louis XIV, King of France, was also cured, he bought Talbor's secret prescription for 3000 gold crowns. Following the healer's death, the formula of his remedy was published. It was quine bark, rose leaves, lemon juice and wine: an English version of the Indian formula.

In contrast to Sir Robert Talbor's glorious end, Spruce returned to England crippled with arthritis. No one was waiting for him in Southampton. He was as poor as when he departed for the Amazon but now unable to work.

This presentation serves then as a remembrance, at least in part, of the glory of the man who lived for science, his country and the well being of humanity.

ACKNOWLEDGEMENTS

The author wishes to thank Dr Ronald Guderian for kindly reviewing the English text; also to Ciba-Geigy Laboratories for a travel grant which enabled the author to present this paper at the Spruce Conference.

NOTES

(1)   M. Jimenez de la Espada, *Relaciones geográficas de Indias*, Madrid, 1881.

(2)   G. Arcos, *Evolución de la medicina en el Ecuador*, 3a. ed., Quito, 1980.

(3)   V. Paredes Borja, *Historia de la medicina en el Ecuador*, Quito, 1963; P. Naranjo, 'Pedro Leiva y el secreto de la quina', *Revista Ecuatoriana de Medicina y Ciencias Biológicas* (1979), **15**, 393.

(4)   J. Jaramillo-Arango, 'Estúdios científicos acerca de los hechos básicos en la historia de la quina', *Revista de la Facultad de Ciencias Médicas, Quito* (1950), **1**, 61.

(5)   H. Gallardo-Moscoso, *Paltas, incas y viracochas. Historia de los vencidos*, Loja, 1970.

(6)   F. Tristram, *Le procès des étoides*, Paris, 1979.

(7)   M. Acosta-Solis, *Investigadores de la geografía y la naturaleza de América tropical: viajeros, cronistas e investigadores con especial referencia al Ecuador*, Quito, 1976.

(8)   V.W. von Hagen, *Grandes Naturalistas en América*, México, 1957.

(9)   Ibid.

(10) M. Alegre-Perez and M.L. Andres, 'Control e informes sanitarios de la Real Botica sobre quina en el periódo ilustrado' and A. Hermosilla, 'La regia Sociedad de Medicina en Sevilla y América en el siglo XVIII', both in *Hispanoamérica y las Academias de Medicina Españolas*. Real Academia Médico y Cirugia de Cádiz, 1992.

(11) R.Spruce, *Notes of a botanist on the Amazon and Andes* (ed. A.R. Wallace), London, 1908, Vol. 2 p. 261.

(12) Ibid. Vol. 2 p. 138-140.

(13) Ibid. Vol. 2 p. 309-310.

(14) Ibid. Vol. 2 p. 204.

(15) UNICEF, 'Malaria', *The Prescriber* No.5, 1993.

14

Orchidaceae Spruceanae: orchids collected by Spruce in South America

Gustavo A. Romero

Orchidaceae Spruceanae: orchids collected by Spruce in South America

Gustavo A. Romero

Harvard University Herbaria, 22 Divinity Avenue,
Cambridge, Massachusetts, U.S.A.

Richard Spruce made over 6,500 numbered collections during his 15-year stay in South America. This paper will be restricted to the orchid numbers he collected (over 180); it was inspired by a small 160-page notebook compiled by Alfred Cogniaux, *Collecteurs Amériques*, 18 x 12 x 1.5 cm. The notebook, part of the Oakes Ames Orchid Library at Harvard University, lists the orchid exsiccata of 99 plant collectors in the New World. Richard Spruce's collections, 124 numbers, occupy three pages. Further herbarium and library work has so far revealed more than 55 additional orchid collections. These 180 orchids include one new genus, 33 new species, and one new variety. Here only Spruce's type specimens are presented. The compilation of all Spruce's orchids in electronic format is an on-going project, and will be published upon completion (a preliminary list is available from the author). Here I will present a log of Spruce's travels emphasizing the orchid species he collected in the upper Rio Negro and Orinoco river basins, as well as the sites he explored in this region. As we will see, Spruce collected not only some of the most showy orchids of the Rio Negro and the Orinoco, but also many rare species, some not collected again for many years, and some others that were neglected and later described from other collections.

Spruce collected live orchids, as announced by W.J. Hooker when he first presented Spruce's trip to the Amazon.[1] In his correspondence with Hooker, Spruce mentions live orchids several times:

> *"I seized the opportunity of the Britannia sailing for London, to tell you how I am getting on ... In the way of living plants I have made a beginning ... I have met with five or six Orchidaceae, the larger ones out of flower; but two small ones in flower, one a pretty Fernandesia, most likely F. lunifera [Lockartia lunifera (Lindl.) Rchb.f.], the other with small sweet-scented yellow flowers".*[2]

> *"Along with these cases you will receive a third, containing chiefly Orchideae, but also a few aroideous and other plants ... I have, besides, a Ward's case full of plants, chiefly Palms, of whose content I will give a list below ...".*[3]

> *"The old low trees along the beach were filled with Orchidaceous plants ... Of the Orchideae only one or two of the smaller species were in bloom; but I gathered of all I could find, and now send them, in order that when they flower you may ascertain whether they include any novelties".*[4]

Later, however, Spruce discarded the idea of sending live plants to England, as shown in the following letter:

> *"The trees ... are generally from their size alone unfitted for cultivation in our conservatories, and the few whose dimensions would not exclude them, are mostly species bearing small and obscurely coloured flowers ... the seeds are so frequently resinous or oily as to lose their vitality when kept long out of the earth and the sending of living plants by a hazardous and tedious* [route]*, partly in small canoes, is an uncertain mode ... coupling these circumstances with my previous very slight acquaintance with the cultivation of plants, your Ladyship will readily understand why my expedition has proved a failure hitherto as regards living plants ...".*[5]

It is not known if any of the living plants that Spruce sent to England during the early stages of his expedition ever survived the transatlantic trip.

Spruce collected about twelve orchid species during his stay in Pará and his ascent of the Amazon river to Manaus. *Galeandra montana* Barb. Rodr. (*Spruce 862*, K!) would have been the only novelty, but his specimen was mis-identified; J. Barbosa Rodrigues described this species in 1881 from one of his own collections.[6] Around Manaus (Barra do Rio Negro in his writings), Spruce collected *Acacallis cyanea* Lindl., one of the showiest orchids of the Amazon and blackwater tributaries of the upper Orinoco (Fig. 1). Lindley's description, in his *Folia Orchidacea*,[7] that *"this is a fine plant"*, is an understatement! Spruce also collected the types of *Habenaria sprucei* Cogn., *H. sylvicultrix* Lindl. ex Kraenzl., *Oncidium sprucei* Lindl., and *Stelis barrensis* Lindl., as well as 11 previously described species, including *Cattleya violacea* (H.B.K.) Rolfe. The latter was first collected by Humboldt and Bonpland some 50 years earlier on the Orinoco (Fig. 2).

Along the Rio Negro and the Uaupés, particularly in Panuré, Spruce collected 26 different orchids, including nine new species. The collections of *Eriopsis sprucei* Rchb.f. (Fig. 1), *Pogonia carnosula* Rchb.f., and *Wullschlaegelia calcarata* Benth. are particularly impressive due to the difficulties Spruce must have endured to procure them. *Eriopsis sprucei* is an epiphytic orchid usually found in large clumps (up to 3 m wide) inhabited by aggressive ants, *Pogonia carnosula* Rchb.f. (now referred to the genus *Triphora*) is so rare it is only known from the type collection, and *Wullschlaegelia calcarata* is a small, inconspicuous saprophytic orchid hardly ever encountered in the field. Spruce also made the first collection of *Bifrenaria petiolaris* (Schltr.) G. Romero & Carnevali,[8] an orchid that so far has defied all attempts of taxonomic classification (Fig. 2).

Spruce continued his journey up the Rio Negro, entering what he called *Terra humboltiana* after crossing the Brazilian-Venezuelan border in Cocui on 3 April 1853.[9] He reached San Carlos de Rio Negro on 11 April 1853. While in San Carlos (and San Felipe, a town across the Rio Negro, now in Colombian territory) Spruce collected several orchids in the surrounding vegetation, including the type of *Pleurothallis spiculifera* Lindl., another miniature plant with flowers 3–4 mm in diameter. From San Carlos, Spruce organized two separate expeditions, one to La Esmeralda in December 1853 and another to Maypures in June 1854.

Spruce first travelled up the Casiquiare, reached the Orinoco, and continued

Fig. 1. Clockwise from top left. *Eriopsis sprucei* Rchb.f. (= *E. sceptrum* Rchb.f. & Warsc.), *Catasetum bergoldianum* Foldats, *Acacallis cyanea* Lindl.

Fig. 2. Clockwise from top left. *Bifrenaria petiolaris* (Schltr.), G. Romero & Carnevali, *Cattleya violacea* (H.B.K.) Rolfe, *Catasetum pileatum* Rchb.f.

Table I. New orchid species collected by Spruce in lowlands of the Amazon and Orinoco river basins and in Peru and Ecuador·

	TOTAL	NEW SPECIES
Amazon and Orinoco	75	20
Peru and Ecuador	105	14
Total	180	34

Total based on preliminary count. The probability of the number of new species being independent of the region of origin is less than 0.05 (G-test with Williams correction).[10]

upstream to arrived at La Esmeralda 24 December 1853, a town Humboldt and Bonpland had visited in May 1800[11] and Robert H. Schomburgk in February 1839.[12] Spruce stayed four days in La Esmeralda, preparing his trip to the Cunucunuma river, and left on 28 December 1853. After a week on the Cunucunuma and unable to go upstream beyond the Tauarupána rapids, Spruce descended to the Orinoco and went back to San Carlos de Rio Negro *via* the Casiquiare. Along the way, he explored the Rio Pacimoni, Caño Baria, and Rio Yatúa, from where he reached the base of Cerro Avispa[13] on 11 February 1854. Spruce collected at least eight orchid species along the Casiquiare and the Pacimoni, including the type specimens of *Elleanthus puber* Cogn. [*Palmorchis puber* (Cogn.) Garay], *Masdevallia sprucei* Rchb.f. and *Stelis viridipurpurea* Lindl. [*S. fraterna* Lindl.], and specimens of *Trisetella triglochin* Rchb.f., a plant not collected again in the Pacimoni until Dunsterville visited the same area in the early 1960s.[14]

When Spruce returned to San Carlos on 24 February 1854, he directed the attention of his expedition to the Maypures rapids. Following Humboldt's route, he left San Carlos on 26 May 1854, went up the Rio Negro and Guainía to Pimichin, crossed overland to Yavita, and went down the Temi to the Atabapo to San Fernando de Atabapo. He left San Fernando and entered the Orinoco on 18 June, reaching Maypures the next day. In Maypures, Spruce collected five orchid species in the riverside vegetation, granite outcrops, and adjacent savannas. The orchids collected included *Catasetum pileatum* Rchb.f.[15] (Fig. 2), *Catasetum bergoldianum* Foldats (Fig. 11), described over a hundred years later from another collection, *Cleistes rosea* Lindl., *Cleistes tenuis* (Rchb.f. ex Griseb.) Schltr., and *Habenaria leprieurii* Rchb.f. Of these, *Cleistes tenuis* has never been collected again in Maypures in either side of the Orinoco river. Spruce fell ill soon after leaving Maypures, and his orchid collections, except for four numbers from the Guainía and the Rio Negro, did not start again until he reached Tarapoto, Peru, in June 1855.

Spruce collected in Peru and Ecuador until 1864, gathering a collection of over 105 orchids, including 14 new species. He collected more orchids in Ecuador and Peru than in the lowlands of the Amazon and Orinoco river basins (105 versus 75; Table 1), but among his collections there were a proportionately larger number of new species from the lowland sites (see Table 1). This is perhaps a reflection of the paucity of early botanical explorations in the lowlands of the Amazon and Orinoco rivers in contrast to the numerous collectors who preceded Spruce in the Andes.

List of new orchid taxa collected by Richard Spruce (1849–1864)

Of Spruce's 34 new taxa (33 species and one variety), John Lindley described 11 new species and one new variety, in his famous *Folia orchidacea*, published between 1852 and 1859. Lindley and Heinrich Gustav Reichenbach described a new genus and species, *Sertifera purpurea*. Professor Reichenbach f.[16] later described eleven new species between 1863 and 1878, Joseph D. Hooker one in 1861, and George Bentham one as part of his famous paper on the Orchidaceae read at the meeting of the Linnean Society of January 1881. Alfred Cogniaux described four as part of the orchid treatment for *Flora Brasiliensis*,[17] Robert A. Rolfe described one in 1896, and Fritz Kränzlin two in 1893 and 1919. Spruce himself proposed a name for one species, *Stelis calodyction* Spruce[18] but it was later described in a separate genus (see *Lepanthes calodyction* Hook.f.; Fig. 9). The most recent species, *Erythrodes sprucei* Garay, was described in 1978.

Acacallis cyanea Lindl., *Folia Orchid.*, *Acacallis* 1. 1853. TYPE: BRAZIL. Barro do Rio Negro, on trees by forest streams, August 1851, *Spruce 1790* (Holotype: ex Herbarium Lindley, K!).
Cogniaux[19] and Prain (1916) cited another specimen with the same number but collected in "sylvis supra arbores secus cataractam Panuré ad Rio Uaupés, July 1853". This second specimen has not been located.
Elleanthus spruceanus Cogn., *Fl. Bras.* 3, 5: 332. 1901. TYPE: PERU. Tarapoto, *Spruce s.n.* (Syntypes: G, ex Herbarium Decandolle & Boissier).
Elleanthus puber Cogn., *Fl. Bras.* 3, 5: 333. 1901. TYPE: VENEZUELA. Amazonas: Río Pacimoni, February 1854, *Spruce 3407* (Lectotype, here designated: ex Lindley Herbarium, K!; Isolectotypes: ex Hooker and Bentham Herbaria, K!).
Cogniaux examined specimens from G, K, and P. Three specimens were located at K (ex Bentham, Hooker, and Lindley Herbaria).
= *Palmorchis puber* (Cogn.) Garay, Caldasia 8: 518, 1962.
Epidendrum spruceanum Lindl. *Folia Orchid. Epidendrum* 80. 1853. TYPE: BRAZIL. Amazonas, Barro do Rio Negro, *Spruce 1466* (Holotype: ex Lindley Herbarium, K!).
= *Epidendrum nocturnum* Jacq. s.l.
Epidendrum variegatum Hook. var. *virens* Lindl. *Folia Orchid. Epidendrdum* 38. 1853. TYPE: BRAZIL. Amazonas, Rio Negro, above Barcellos, Spruce 1948 (Lectotype, here selected, ex Lindley herbarium, K!; Isolectotype, ex Bentham Herbarium, K!).
Lindley cited specimens from Loddiges, Spruce, and Funck and Schlim without selecting a type.
= *Encyclia vespa* (Vell.) dressler.
Eriopsis sprucei Rchb.f., *Ann. Bot. Syst.* 6: 663. 1863. TYPE: BRAZIL. Rio Negro, June 1852, *Spruce 2390* (Holotype: Reichenbach Herbarium 37992, W!; Isotype: ex Bentham Herbarium, K!).
There are at least two sheets in the Reichenbach herbarium with the number *Spruce 2390*, one from "Gapó of Rio Negro, from Manaváca upwards" (Reichenbach Herbarium 25907), the second from "Ad flum Guainia supor ostium flumini Casiquiare" (Reichenbach Herbarium 37992). The locality reported by

Reichenbach f. in the protologue ("Rio Negro") provides an unambiguous criterion for the selection of the holotype.

Generally referred to *E. sceptrum* Rchb.f. & Warsc.

Erythrodes sprucei Garay, *Fl. Ecuador* 9: 284. 1978. TYPE: ECUADOR. Chimborazo: Río Chasuán, July 1860, *Spruce 6138* (Holotype: Reichenbach Herbarium 37767, W!; Isotype: ex Hooker Herbarium, K!).

Habenaria sprucei Cogn. Fl. Bras. 3, 4: 40. 1893. TYPE: BRAZIL. Amazonas, Manaus, *Spruce 1221* (Holotype: BR).

Habenaria sylvicultrix Lindl. ex Kraenzl., *Bot. Jahr. Syst.* 16: 101. 1893. TYPE: BRAZIL. Amazonas, Manaus, *Spruce 1262* (Lectotype, here designated: ex Lindley Herbarium, K!; Isolectotype: BM).

Kränzlin cited specimens from K and BM without selecting a type.

Lepanthes calodictyon Hook.f., *Bog. Mag.* 137: t. 5259. 1861. TYPE: ECUADOR. Spruce s.n. (Holotype: K).

Spruce first assigned this species to *Stelis*: "As respect to foliage, a fairy *Stelis* (*S. calodyction*, Mss.) with roundish pale green leaves, beautifully reticulated with the purple veins, far excels every other plant seen in the Cinchona woods"[20] (see Fig. 9).

Lepanthes roseola Rchb.f. *Linnaea* 41: 46. 1877. TYPE: ECUADOR. Quito, *Spruce 5954* (Holotype: Reichenbach Herbarium 54529, W!).

Masdevallia sprucei Rchb.f., *Otia Bot. Hamburg.* 1: 17. 1878. TYPE: VENEZUELA. Amazonas, Río Uaienaka, tributary of Río Pacimoni, February 1854, *R. Spruce 3369* (Holotype: Lindley Herbarium, K!, fragment of holotype, Reichenbach Herbarium 38718, W!; Isotype: ex Bentham Herbarium, K!).

Reported from "... lignis ad rivulum Uaienaka fluvii Parimoni tribut", a locality often referred to Brazil[21].

Maxillaria nitidula Rchb.f., *Linnaea* 41: 30. 1877. TYPE: ECUADOR. Río Pastasa, Puente de Agoyán, February 1859, *Spruce 5896* (Holotype: not located; Isotypes: ex Bentham Herbarium, K!, ex Hooker Herbarium, K!).

Oncidium pastasae Rchb.f., *Linnaea* 41: 21. 1877. TYPE: ECUADOR. Quito, "secus fluvium Pastasa", October 1857, *Spruce 5078* (Holotype: Reichenbach Herbarium 5058, W!; Isotypes: ex Bentham Herbarium, K!, Reichenbach Herbarium 4478, W!).

Oncidium sprucei Lindl. *Folia Orchid., Oncidium* 14. 1855. TYPE: BRAZIL. Common on trees at the junction of Rio Negro and Solimoes, June 1851, *Spruce 1526* (Lectotype: here designated, ex Lindley Herbarium, K!; Isolectotypes: GH!, ex Bentham Herbarium, K!, Reichenbach Herbarium 27559, W!).

Lindley cited two specimens in the protologue, *Gardner 2732* and *Spruce 1526* without selecting a type.

Oncidium umbonatum Rchb.f. *Linnaea* 41: 24. 1877. TYPE: ECUADOR. Quito, August 1859, *Spruce 6074* (Holotype: Reichenbach Herbarium 6483, W!; Isotype: Reichenbach Herbarium 6484, W!).

Ornithidium cordyline Rchb.f. *Linnaea* 41: 34. 1877. TYPE: ECUADOR. Quito, Tunguragua, January 1861, *Spruce 6242* (Holotype: Reichenbach Herbarium 39817, W!).

Ornithidium quitoense Rchb.f. *Linnaea* 41: 34. 1877. TYPE: ECUADOR. Quito, Tunguragua, January 1861, *Spruce 6244* (Holotype: Reichenbach Herbarium 39815, W!).

Physurus stigmatopterus Rchb.f., *Xenia Orchid.* 2: 185. 1874. TYPE: BRAZIL. Amazonas, Rio Uaupés, Panuré, October 1852, *Spruce 2660* (Holotype: Lindley Herbarium, K! *fide* annotation of L.A. Garay, 1977; Isotype: ex Bentham Herbarium, K!).

= *Erythrodes stigmatoptera* (Rchb.f.) Pabst, *Orquídea* (*Rio de Janeiro*) 18: 215. 1956.

Pleurothallis acutissima Lindl. *Folia Orchid., Pleurothallis* 43. 1859. TYPE: BRAZIL. Amazonas, Rio Negro, caatingas about S. Gabriel, on the bark of trees, May 1852, *Spruce 2302* (Holotype: ex Lindley Herbarium, K!).

Pleurothallis fimbriata Lindl., *Folia Orchid.*, *Pleurothallis* 43. 1859. TYPE: BRAZIL. Amazonas, Panuré, caatingas, creeping on trees, October 1852 (September in fieldbook), *Spruce 2459* (Holotype: ex Lindley Herbarium, K!).
Generally referred to *Pleurothallis miqueliana* Lindl. (Fig. 10).

Pleurothallis spiculifera Lindl. *Folia Orchid.*, *Pleurothallis* 43. 1859. TYPE: VENEZUELA. Amazonas, San Carlos de Río Negro, October 1853, *Spruce 3154* (Holotype: ex Lindley Herbarium, K!).

Pleurothallis serrifera Lindl., *Folia Orchid.*, *Pleurothallis* 34. 1859. TYPE: BRAZIL. Amazonas, Panuré, caatingas, December 1852, *Spruce 2724* (Holotype: ex Lindley Herbarium, K!).

Pleurothallis sprucei Lindl., *Folia Orchid.*, *Pleurothallis* 35. 1859. TYPE: BRAZIL. "Amazonas, caatingas by the river Uaupés, on trees", December 1852, *Spruce 2725* (Holotype: ex Lindley Herbarium, K!).

Pogonia carnosula Rchb.f., *Nederl. Kruidk. Arch.* 4: 323. 1859. TYPE: BRAZIL. Uaupés, February 1853, *Spruce 2902* (Holotype: K, ex Lindley Herbarium!; Isotypes: ex Bentham and Hooker Herbaria, K!).
= *Triphora carnosula* (Rchb.f.) Schltr., *Beih. Bot. Centralblatt* 42: 76, 1925.

Ponthieva sprucei Cogn., *Fl. Bras.* 3, 4: 274. 1895. TYPE: PERU. Tarapoto, *Spruce 3936* (Syntypes: G, K; Isosyntype: GH!).
Cogniaux cited specimens from G and K without selecting a type.

Ponthieva andicola Rchb.f. *Linnaea* 41: 52. 1876. TYPE: ECUADOR. Llalla, August 1859, *Spruce 5998* (Holotype: Herbarium Reichenbach 50378, W!; Isotype: GH!).

Sertifera purpurea Lindl. & Rchb.f., *Linnaea* 41: 63. 1877. TYPE: ECUADOR. Tunguragua, May 1858, *Spruce 5394* (Lectotype, selected here: ex Lindley Herbarium, K!; Isolectotypes: Reichenbach Herbarium 47916, W!, Ex Bentham and Hooker Herbaria, K!).

Stelis barrensis Lindl., *Folia Orchid.*, *Stelis* 6. 1858. TYPE: BRAZIL. Rio Negro, Barra, shady forest on trees, July 1851, *Spruce 1656* (Holotype: ex Lindley Herbarium, K!).
= *Stelis papaquerensis* Rchb.f. (*fide* Garay)[22].

Stelis calodyction Spruce, *nom. nud.*[23]; see *Lepanthes calodictyon* Hook.f.

Stelis filiformis Lindl., *Folia Orchid.*, *Stelis* 6. 1858. TYPE: VENEZUELA. Amazonas, Río Casiquiare, on trees on the banks of Lake Vasiva, February 1854, *Spruce 3438* (Holotype: ex Lindley Herbarium, K!; Isotype: P!).

Stelis viridi-purpurea Lindl., *Folia Orchid.*, *Stelis* 3. 1858. TYPE: VENEZUELA. Amazonas, Río Casiquiare, Uaiavaka, February 1854, *Spruce 3368* (Holotype: ex Lindley Herbarium, K!; Isotype: P!).
Stelis fraterna Lindl. (fide Garay)[24].

Telipogon pogonostalix Rchb.f., *Linnaea* 41: 72. 1877. TYPE: ECUADOR. Chimborazo, San Antonio, June–September 1860, *Spruce 6135* (Holotype: Reichenbach Herbarium 30096, W!).
= *Stellilabium pogonstalix* Garay & Dunsterv., *Venez. Orch. Illustr.* 2: 336. 1961.

Telipogon sprucei Kraenzl., *Ann. Naturhist. Hofmus.* 33: 23. 1919. TYPE: ECUADOR. Quito, Llalla, August 1859, *Spruce 6076* (Syntypes: Reichenbach Herbarium 30084, W!).
There are two different specimens of *Spruce 6076* in the cited sheet.

Vanilla sprucei Rolfe, *J. Linn. Soc., Bot.* 32: 461. 1896. TYPE: BRAZIL. "In shady woods near the Uaupés River", December 1852, *Spruce 2727* (Holotype: ex Lindley Herbarium, K!).

Wullschlaegelia calcarata Benth., *J. Linn. Soc., Bot.* 18: 342. 1881. TYPE: BRAZIL. Panuré and Río Uaupés, January 1853, *Spruce 2847* (Holotype: Ex Bentham Herbarium, K!).

ACKNOWLEDGEMENTS

I am grateful to the curators and staff of BR, P, W and particularly the Orchid Herbarium at K, for the loan of critical herbarium specimens and access to library material, and to R.E. Schultes for depositing his extensive collection of Spruce-related documents at the Botany Library, Harvard University. I thank The Harvard University Herbaria and the American Orchid Society for providing financial support, and S. Madriñán, P.L. Beer-Romero and P. Stern for commenting on the manuscript.

NOTES

(1) "Mr Spruce ... desires to supply cultivators with seeds and living plants, according to terms hereafter to be agreed upon ..." in J.D. Hooker, 'Mr. Spruce's intended voyage to the Amazon river', *Hooker's Journal of Botany and Kew Garden Miscellany* (1849), **1**, 21.

(2) Letter to W.J. Hooker, dated Pará, 3 August 1849, *Hooker's Journal of Botany and Kew Garden Miscellany* (1849), **1**, 346.

(3) Letter to W.J. Hooker, dated Pará, 7 October 1849, *Hooker's Journal of Botany and Kew Garden Miscellany* (1850), **2**, 65.

(4) Ibid., 67.

(5) From 'Extract of a letter from Mr Spruce to the Countess of Carlisle' dated 1854, copy at the Botany Library, Harvard University.

(6) J. Barbosa Rodrigues, *'Galeandra montana'*, *Revista de Engenharia* (1881), **3**, 73.

(7) J. Lindley, *Folia Orchidacea, Acacallis*, London, 1853, 1.

(8) *Bifrenaria petiolaris* (Schltr.) G. Romero & Carnevali, *comb. nov.* Basionym: *Maxillaria petiolaris* Schltr., *Beih. Bot. Centralbl.* 42: 133. 1925. Spruce's specimen (*Spruce 2724a*) apparently lost its flowers in transit (Spruce did describe flower colour in his label) and it remained unidentified at K until the author recognized it in September 1993. The species was first described in 1925 as *Maxillaria petiolaris* Schltr. based on two G. Hübner's specimens. Hoehne, citing an earlier binomial (*Maxillaria petiolaris* A. Rich.), proposed a new name, *Maxillaria rudolfii* Hoehne(a). It was then re-described in 1958 as *Bifrenaria minuta* Garay from a G.C.K. Dunsterville's specimen(b), but later transferred to *Maxillaria* (as *M. perparva* Garay & Dunsterv; (c)). We (Carnevali and Romero, in (d)) made the combination in *Bifrenaria* using Hoehne's epithet, not realizing the A. Richard's binomial had never been validly published (Reichenbach cited it in the synonymy of his *Maxillaria desvauxiana*(e)). On final analysis, this most likely belongs in an undescribed genus of the *Bifrenaria* Lindl. group of Lycastinae! ((a)F.C. Hoehne, 'Reajustamento de algumas espécies de Maxillarieas do Brasil, con a criação de dois novos gêneros para elas', *Arquivos de Botânica do Estado de S. Paulo* (1947), **2(4)**, 73; (b)L.A. Garay, Studies in American orchids. *Botanical Museum Leaflets Harvard University* (1958), **18**, 206; (c)G.C.K. Dunsterville and L.A. Garay, *Venezuelan Orchids Illustrated* London, 1976, vol. VI, 37; (d)L. Brako and J.L. Zarucchi, 'Catalogue of the flowering plants and Gymnosperms of Peru', *Monographs in Systematic Botany from the Missouri Botanical Garden* (1993), **45**, 766; (e) H.G. Reichenbach f., *Maxillaria desvauxiana, Refugium Botanicum* (1882), **2**, t. 134.)

(9) R. Spruce, *Notes of a botanist on the Amazon and Andes* (ed. A.R. Wallace), London, 1908, vol. 1, 346.

(10) R.R. Sokal and F.J. Rohlf, *Biometry*, San Francisco. 1981, 736–738.

(11) O. Huber and J.J. Wurdack, 'History of botanical exploration in Territorio Federal Amazonas, Venezuela', *Smithsonian Contributions to Botany* (1984), **56**, 44.

(12) R.H. Schomburgk, *Travels in Guiana and on the Orinoco during the years 1835–1839*, Georgetown, British Guiana, 1931, 170; Huber and Wurdack op. cit. (11), 57.

(13) Cerro Avispa was not visited again by botanists until 1959 when an expedition from the New York Botanical Garden collected in the lower part of the mountain as part of the Botany of the Guayana Highlands project (B. Maguire, J.J. Wurdack, et al., The botany of the Guayana highlands - Part V. *Memoirs of the New York Botanical Garden* (1964), **10(5)**, 1)

(14) G.C.K. Dunsterville and L.A. Garay, *Venezuelan Orchids Illustrated*, London, 1972, vol. V, 168.

(15) Spruce's specimen of *C. pileatum* remained misfiled at Kew for over 30 years (a). Had someone described this species soon after the collections arrived in Kew, a major calamity in the field of orchidology could have been averted! *Catasetum pileatum* Rchb.f. was first described in 1882 from dry flowers sent by Lindens, who ran the famous *L'Horticulture International* of Brussels. Richard Payer, one of Linden's collectors, collected live plants along the Orinoco river in 1881. This shipment of plants apparently did not flourish in Brussels and the species was forgotten until it was re-described in 1886 from a drawing by E. Bungeroth as *Catasetum bungerothii* N.E. Brown. Bungeroth was another of Linden's collectors who re-discovered the species in the upper Orinoco river in 1886. The new introduction was well received in Europe: "The appearance of this Orchid in a flowering state at Steven's Room on December 16 [1886] last caused a flutter among orchidologists ... At the sale in question the best plant realised fifty guineas ..." (b). A controversy, however, followed over whether the proper name was *Catasetum pileatum* or *C. bungerothii*. Reichenbach argued that *C. bungerothii* was the *C. pileatum* he described in 1882 (c), while the Lindens contended that they had never sent flowers of this showy species to Reichenbach ("I cannot remember what *Catasetum* we may have sent to the Professor, but I am quite sure it was not *C. Bungerothi*. This species is so extraordinary, that my father and myself must have recollected it"; (d). This heated exchange took place during the last few years of Reichenbach's fruitful life and was arguably a major factor in his decision to lock his herbarium for 25 years after his death in May 1889," ... an action which staggered the scientific world, and struck a cruel blow at the progress of orchidology" (e) ((a) R.A. Rolfe, '*Catasetum bungerothii* var. *aureum*', *Gardeners' Chronicle* Ser. 3, (1889), **6**, 466; R.A. Rolfe, '*Catasetum pileatum*', *Orchid Review* (1915), **23**, 296; (b) in M.T. Masters, *Catasetum bungerothii*. *Gardeners' Chronicle* Ser 3, (1887), **1**, 139; (c) e.g. F. Sander, '*Catasetum pileatum*', in *Reichenbachia, orchids illustrated and described* (F. Sander), St Albans, 1890, vol. 2, 92; (d) in L. Linden, *Catasetum bungerothii*. *Gardeners' Chronicle* Ser 3, (1890), **7**, 618, see also L. Linden, L'introduction du *Catasetum Bungerothi*. *Le Journal des Orchidées* (1890), **1**, 168–170, 184–186; (e) R.A. Rolfe, 'The Reichenbergian Herbarium', *Orchid Review* (1913), **21**, 273).

(16) Reichenbach f. purchased at least a partial set of Spruce's Amazonian collections and a complete set of his Andean plants. Spruce, in a letter to George

Bentham (Hurstpierpoint, 6 November 1865) described Reichenbach's reaction to the Andean collections: "I did not expect Reichenbach would want any more of my plants. I sold him my own set of Equatorian plants, and I spent many weary hours in writing out for him notes on the colours of the flower in the living plant; in return he wrote me a letter of 4 closely-written quarto pages full of scolding & grumbling. I have not to this day read more than the first page of it ...". Spruce and Reichenbach continued to exchange correspondence, and Reichenbach later published a complete description of the geographical distribution of Spruce's South American plant collections (H.G. Reichenbach f., 'Zum geographischen Verstandniss der Amerikanischen Reisepflanzen des Herrn Dr. Spruce', *Botanische Zeitung* (1873), **31**, 28–29).

(17)   A. Cogniaux, 'Orchidaceae', *Flora Brasiliensis* (ed. K.F.P. von Martius), Munich, Leipzig, 1893–1906, vol. 3, parts 4–6.

(18)   R. Spruce, *Report on the expedition to procure seeds and plants of the* Cinchona succirubra, *or Red Bark Tree*, London, 1861, 43.

(19)   Cogniaux, op. cit. (17), part 5, 525.

(20   Spruce, op.cit.(18), 43.

(21)   e.g. Cogniaux, op. cit.(17), part 4, 329.

(22)   L.A. Garay, 'Systematics of the Genus *Stelis* Sw.', *Botanical Museum Leaflets Harvard University* (1980) **27**, 192.

(23)   Spruce, op. cit. (18), 43; op. cit. (9) vol 2, 278.

(24)   Garay, op. cit. (22), 204.

FURTHER REFERENCES CONSULTED

L.A. Garay, 'Orchidaceae', in *Flora of Ecuador* (eds G. Harling and B. Sparre), Stockholm, 1978, no. 9.

R.A. Rolfe, 'Dr. Lindley', *Orchid Review* (1917), **25**, 75-79.

F.A. Stafleu and R.S. Cowan, *Taxonomic Literature* (2nd ed.), Utrecht, 1976, 1979, vols. 1-2.

15

The biology of mamure *Heteropsis spruceana* Schott (Araceae)

Gustavo A. Romero and Paul E. Berry

The biology of mamure *Heteropsis spruceana* Schott (Araceae)

Gustavo A. Romero[1] and Paul E. Berry[2]

[1]*Harvard University Herbaria, Divinity Ave., Cambridge, Massachusetts*
[2]*Missouri Botanical Garden, Tower Grove Avenue, St. Louis, Missouri, USA*

Heteropsis spruceana Schott ("mamure" or "tripa de pollo" in Venezuela, "cipó titica" in Brazil, "tamishi delgado" or "támishe" in Peru) was first collected by Richard Spruce in S. Gabriel da Cachoeira, Brazil, in 1852. The plant produces aerial roots up to 30 m long and 4-9 mm in diameter. These roots are woody, tough, and flexible. Indian groups in the Amazon and Orinoco river basins have for centuries used them as rope for house construction and tying, and, after cleaning and splitting, for weaving baskets and fish traps.

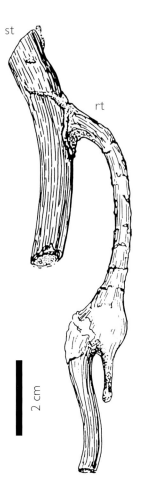

Furniture makers in Venezuela have recently discovered that the roots of *Heteropsis* are a reasonable substitute, both in price and quality, for osier (*Salix* spp., Salicaceae, "mimbre" in Spanish), an expensive import. As a result, the roots of *Heteropsis* are currently being harvested indiscriminately and the plant has disappeared from many areas in the Venezuelan Amazon. *Heteropsis* is potentially an ideal non-timber forest product. There is, however, little information available on its reproductive biology, growth rates, the effect of root-removal on the long-term survival of the plant, and root regeneration.

We present preliminary data from four field trips between January 1991 and July 1993 in southern Venezuela. Seeds of *Heteropsis* germinate on the ground. Seedlings grow along the forest floor until they find and climb a host tree. Climbing plants develop side branches and aerial roots after attaining a height of 2-3 m above the forest floor. This change in growth form is apparently due to changes in light level in the canopy. Plants that climb small trees or shrubs will descend to the ground and grow along the forest floor until finding another tree.

Fig. 1. Node in root of *Heteropsis spruceana*. Root (rt) attached to a section of stem (st). Small, stunted growth on the right was parasitized earlier; thicker root on the left is actively growing (scale bar = 2 cm).

Plant growth rates increase after reaching 2-3 metres in height. The aerial roots grow up to five metres per year. As plants continue to climb, the stem rots away close to the ground and plants then rely exclusively on the aerial roots for nourishment. The tip of the roots are attacked by moth larvae (Noctuidae) that apparently feed on meristematic tissue. The plant responds by producing a bulge or node above the injured tip with a new lateral root tip (Fig. 1). Depending on the rate of predation, the distance between root nodes (the internodes) varies from a few centimetres to 15 metres long. Furniture makers prefer roots with relatively long internodes (< 1 m), but these constitute only 20-35% of the roots surveyed in the experimental plot. Root-removal experiments indicate that the survival rate of plants that lost all their roots is 5% whereas survival is 100% for plants that only lost half their roots.

Based on these data, we recommend that collecting permits be issued only to local people who use the roots to make furniture and/or other crafts. These individuals tend to gather only high quality roots (i.e. roots with long internodes), leaving behind sufficient roots to sustain the plant, and thus assuring the long-term, sustainable harvest of roots. Other individuals less experienced with *Heteropsis*-craft should be issued permits accompanied by a pamphlet that explains the importance of leaving enough roots for sustained production.

ACKNOWLEDGEMENTS

We thank the Heinz Foundation, the C. Lindbergh Fund, the Harvard University Herbaria, and the Missouri Botanical Garden for financial support, and Venezuela's MARNR, SADA-Amazonas, FONAIPAP/Amazonas, and Fundación Instituto Botánico for logistic support.

16

Richard Spruce and the palms of the Amazon and Andes

Andrew Henderson

Richard Spruce and the palms of the Amazon and Andes

Andrew Henderson

New York Botanical Garden, Bronx, NY 10458, USA

Introduction

Richard Spruce (1817–1893) travelled and collected plants in the Amazon and Andes of South America in the years 1849–1864. Amongst his collections were many palms, and on his return to England he published an important work on this family.[1] This chapter reviews the state of knowledge of Amazon and Andean palm taxonomy prior to Spruce's journey; his collections in the Amazon and Andes; his interaction with Alfred Russel Wallace; and his contribution to our knowledge of Amazon and Andean palms.

Amazon and Andean palm taxonomy before 1850

Taxonomic knowledge of Amazon palms was well advanced before Spruce's work through the work of the remarkable German botanist Carl von Martius (1794–1868). Martius arrived in Rio de Janeiro, Brazil, in 1817 when he was 23 years old, accompanying an Austrian scientific expedition.[2] With the zoologist Johann Baptist von Spix, he travelled overland from Rio de Janeiro to the mouth of the Amazon, arriving in Belém in 1819. They spent the next 11 months collecting along the river,[3] travelling from Belém to Manaus where they parted company. Spix chose to go up the Rio Negro while Martius continued west along the Amazon and then northwest up the Rio Japurá. He reached Araracuara in Colombia on the Rio Caquetá (the Colombian name for the Rio Japurá) before turning back. Martius collected numerous palms, which were his principal interest. On his return to Munich he began his monograph on palms, perhaps the greatest monograph ever published on any family of flowering plants: his three volume, large folio *Historia naturalis palmarum*.[4] Martius treated his Amazon palm collections in volume two. Each species was fully described and beautifully illustrated, both with habit and details of flowers and fruits. A total of 85 species from the Amazon part of his journey were described, 54 of which are currently recognized.[5]

The taxonomy of Andean palms was not as well known prior to 1850. During the years 1778–1788 the Spanish botanists Hipólito Ruíz and José Pavón collected in Peru and Chile.[6] From Peru, on eastern Andean slopes, mostly in the Departments of Huánuco, Junín and Pasco, they described several new genera of palms, including *Iriartea* and *Phytelephas*, and 10 new species. Shortly after this, in 1799–1804, Alexander von Humboldt and Aimé Bonpland travelled in the Andes, and just reached the northern part of the Amazon region in Venezuela.[7] Subsequently Kunth[8] described three new genera and one species of Amazon palm and about 10 species of Andean palms. Thus at the time of Spruce's work in the Andes only about 20 species of palms were described from the region. Currently 86 species in 21 genera are recognized from the Andes above 1,000 m elevation.[9]

Spruce's palm collections

Spruce arrived in the city of Belém (formerly Pará) in 1849 when he was 32 years old. It seems that he had a rather ambivalent attitude toward palms. He was well aware of their importance to indigenous people in the Amazon, and made many observations on their usefulness,[10] and also collected artefacts made from palms. At the same time he was aware that many species had already been collected and described by Martius.[11] Since the main purpose of Spruce's trip was to collect novelties, he initially collected few specimens of palms. He generally avoided larger species because of the difficulty of collecting, drying and shipping them, and he only collected smaller species when time allowed.[12]

In his account of Amazon palms, Spruce[13] detailed the localities in which he collected palms on his journey. He also kept a separate notebook on his palm collections, but unfortunately this has not been located. However, using his descriptions and herbarium specimens one can give an approximation of the number of collections in each locality (Table 1). Spruce collected about 67 palm specimens (Table 2). These were not numbered chronologically, and apparently the numbers were added later. It is possible that not all collections are accounted for; some numbers are missing and some are repeated.

On his return to England from South America in 1864, Spruce worked on his collections in his house in Yorkshire, where they had been sent by Joseph Hooker of Kew. The result of this work is one of the most important papers on neotropical palm taxonomy.[14] From Spruce's first five localities, i.e., those in the Amazon region, he described 47 new species from at least 62 collections. Of these, 10 are currently accepted at either specific or varietal rank.[15] These are: *Attalea racemosa* Spruce, *Bactris balanophora* Spruce, *B. bidentula* Spruce, *B. concinna* Mart. var. *inundata* Spruce, *Geonoma aspidifolia* Spruce, *G. maxima* (Poit.) Kunth var. *chelidonura* (Spruce) Henderson, *Lepidocaryum tenue* Mart. var. *casiquiarense* (Spruce) Henderson, *Syagrus inajai* (Spruce) Becc., *S. orinocensis* (Spruce) Burret and *Wettinia maynensis* Spruce. From his last two localities, i.e., those in the Andes, he described one new species, *Phytelephas aequatorialis*, from approximately five collections.

Table 1. Spruce's collecting localities for palms.

LOCALITY	NUMBER OF COLLECTIONS
1. Pará (i.e., Belém and vicinity)	5
2. Amazon, between the Tapajós and Trombetas (i.e., Santarém and vicinity)	3
3. Confluence of Rio Negro and Amazon (i.e., Manaus and vicinity)	17
4. Upper Rio Negro, Uaupés, Casiquiare, Orinoco	32
5. Tarapoto, in the Andes of Maynas	5
6. Canelos, eastern side of Andes	?4
7. Guayaquil, western side of Andes	1

Table 2. Spruce's palm collections.

Spec. No.	Spruce's Name	Currently Accepted Name (if different from Spruce's)	Locality Number
3	*Oenocarpus minor* Mart.		3
4	*Iriartea setigera* Mart.	*Iriartella setigera* (Mart.) H. Wendl.	3
5	*Euterpe mollissima* (K)	*Euterpe catinga* Wallace	3
6	*Maximiliana inajai* (K)	*Syagrus inajai* (Spruce) Becc.	3
8	*Bactris concinna* Mart.		3
8a	*Bactris concinna* var. *inundata* (K)		3
9	*Bactris bidentula* (K)		3
15	*Bactris hylophila* (K)	*Bactris hirta* Mart. var. *hirta*	3
16	*Geonoma pauciflora* Mart.	?	3
17	*Bactris negrensis* (K)	*Bactris simplicifrons* Mart.	3
18	*Geonoma tuberculata* (K)	*Geonoma maxima* (Poit.) Kunth var. *chelidonura* (Spruce) Henderson	3
21	*Euterpe oleracea* Mart.	*Euterpe precatoria* Mart.	3
22	*Bactris bifida* Mart.		3
23	*Mauritia quadripartita* (K)	*Lepidocaryum tenue* Mart. var. *tenue*	4
24	*Leopoldinia major* Wallace		4
26	*Iriartea ventricosa* Mart.	*Iriartea deltoidea* Ruíz. & Pav.	5
28	*Geonoma microspatha* (K)	*Geonoma deversa* (Poit.) Kunth	4
29	*Geonoma hexasticha* (K)	*Geonoma maxima* (Poit.) Kunth var. *maxima*	4
30	*Geonoma densiflora*	*Geonoma maxima* (Poit.) Kunth var. *chelidonura* (Spruce) Henderson	4
30*	*Geonoma discolor* (K)	*Geonoma maxima* (Poit.) Kunth var. *maxima*	2
31	*Bactris microcarpa* (K)	*Bactris hirta* Mart. var. *hirta*	4
31*	*Astrocaryum munbaca* Mart.	*Astrocaryum gynacanthum* Mart.	?
32	*Geonoma paniculigera* Mart.	*Geonoma deversa* (Poit.) Kunth	4
33	*Geonoma densiflora* var. *monticola*	*Geonoma maxima* (Poit.) Kunth var. *chelidonura* (Spruce) Henderson	4
34	*Geonoma personata* (K)	*Geonoma maxima* (Poit.) Kunth var. *chelidonura* (Spruce) Henderson	4
35	*Bactris brevifolia* (K)	*Bactris simplicifrons* Mart.	4
36	*Geonoma discolor* (K)	*Geonoma maxima* (Poit.) Kunth var. *maxima*	2
37	*Bactris floccosa* (K)	*Bactris cuspidata* Mart.	2
38	*Astrocaryum munbaca* Mart.	*Astrocaryum gynacanthum* Mart.	?
39	*Mauritia subinermis* (K)	*Mauritiella armata* (Mart.) Burret	4
40	*Mauritia casiquiarensis* (K, NY)	*Lepidocaryum tenue* var. *casiquiarense* (Spruce) Henderson	4

SPEC. NO.	SPRUCE'S NAME	CURRENTLY ACCEPTED NAME (IF DIFFERENT FROM SPRUCE'S)	LOCALITY NUMBER
40	*Mauritia guainiensis* (K)	*Lepidocaryum tenue* var. *casiquiarense* (Spruce) Henderson	4
41	*Geonoma microspatha* var. *pacimonensis* (K)	*Geonoma deversa* (Poit.) Kunth	4
42	*Geonoma macrospatha* (K)	*Geonoma baculifera* (Poit.) Kunth	4
43	*Geonoma chelidonura* (K)	*Geonoma maxima* (Poit.) Kunth var. *chelidonura* (Spruce) Henderson	4
43	*Attalea humboldtiana* (K)	*Attalea butyracea* (Mutis) Wess. Boer	4
44	*Geonoma chelidonura*	*Geonoma maxima* (Poit.) Kunth var. *chelidonura* (Spruce) Henderson	4
45	*Euterpe catinga* Wallace		3
46	*Desmoncus riparius* (K)	*Desmoncus polyacanthos* Mart. var. *polyacanthos*	4
47	*Astrocaryum acaule* Mart.		4
49	*Cocos orinocensis* (K)	*Syagrus orinocensis* (Spruce) Burret	4
50	*Leopoldinia piassaba* Wallace		4
51	*Bactris turbinata* (K)	*Bactris hirta* Mart. var. *hirta*	4
52	*Bactris carolensis* (K, P)	*Bactris simplicifrons* Mart.	4
53	*Bactris balanophora* (K)		4
54	*Attalea racemosa* (K)		4
55	*Mauritia aculeata* Kunth	*Mauritiella aculeata* (Kunth) Burret	4
56	*Mauritia carana* Wallace		?
57	*Oenocarpus minor* Mart.		3
58	*Morenia poeppigiana* Mart.	*Chamaedorea linearis* (Ruíz & Pav.) Mart.	?
61	*Phytelephas macrocarpa* Ruíz & Pav.	*Phytelephas macrocarpa* subsp. *macrocarpa* (Ruíz & Pav.) Barfod	5
63	*Oenocarpus multicaulis* (K)	*Oenocarpus mapora* H. Karst.	5
64	*Phytelephas aequatorialis* (K)		7
65	*Nunnezharia fragrans* Ruíz and Pav.	*Chamaedorea fragrans* (Ruíz & Pav.) Mart.	5
67	*Nunnezharia geonomoides* (K)	*Chamaedorea pinnatifrons* (Jacq.) Oerst.	5
69	*Geonoma paraensis* (K)	*Geonoma maxima* (Poit.) Kunth var. *maxima*	1
70	*Geonoma negrensis* (K)	*Geonoma maxima* (Poit.) Kunth var. *maxima*	4
70	*Bactris negrensis* var. *minor*	*Bactris simplicifrons* Mart.	3
71	*Geonoma baculifera* Poit.	*Geonoma baculifera* (Poit.) Kunth	1
71a	*Geonoma baculifera* Poit.	*Geonoma baculifera* (Poit.) Kunth	1
71b	*Geonoma baculifera* Poit.	*Geonoma baculifera* (Poit.) Kunth	1

Spec. No.	Spruce's Name	Currently Accepted Name (if different from Spruce's)	Locality Number
73	*Geonoma chelidonura* (K)	*Geonoma maxima* (Poit.) Kunth var. *chelidonura* (Spruce) Henderson	4
75	*Geonoma aspidiifolia* (K)		3
77	*Bactris uaupensis* (K)	*Bactris simplicifrons* Mart.	4
78	*Bactris tenuis* Wallace	*Bactris simplicifrons* Mart.	4
80	*Bactris brevifolia* (K)	*Bactris simplicifrons* Mart.	4
81	*Bactris bicuspidata* (K)	*Bactris acanthocarpa* Mart. var. *acanthocarpa*	1
82	*Desmoncus macroacanthos* Mart.	*Desmoncus phoenicocarpus* Barb. Rodr.	1
83*	*Maximiliana inajai*	*Syagrus inajai* (Spruce) Becc.	3
90*	*Euterpe mollissima*	*Euterpe catinga* Wallace	3
?	*Wettinia maynensis*		5
?	*Bactris simplicifrons* Mart.		3

Specimen numbers are either from the actual specimen or from Spruce,[16] where they are given as "S. hb. Palm" or "Mus. Kewense" (these latter are marked with an asterisk). All names are Spruce's except where indicated, and types are indicated by the herbaria in which they are kept (K = Kew; NY = New York; P = Paris). Currently accepted names are from Henderson.[17] The locality number refers to those in Table 1.

Spruce and Alfred Russel Wallace

Alfred Russel Wallace (1823–1913) was 25 years old (7 years younger than Spruce), when he began his Amazon journey in 1848. He arrived in Belém a year before Spruce. Though primarily an entomologist, Wallace was interested in all aspects of Amazon life, and collected various natural history specimens.[18] Wallace and Spruce met for the first time in Brazil in Santarém in late October or early November 1849. From there Wallace travelled on to Manaus while Spruce stayed on in Santarém, not arriving in Manaus until December 1850. Soon after Wallace arrived in Manaus he began his first trip up the Rio Negro. The next time the two naturalists met was on or soon after 15 September 1851 when Wallace returned from the Rio Negro; by then Spruce had been in Manaus for at least 9 months. This was an important meeting. Wallace was one of the first botanical collectors to travel on the black water Rio Negro (Martius had stayed on the white water rivers, and even though Spix had gone up the Rio Negro he did not collect palms). Several quite different species of palms, undescribed at that time, occur in this area, and Wallace had described and sketched them. He had thus made something of a botanical scoop. When he met Spruce in Manaus he showed him the sketches of the new palms. Balick[19] cites a later letter from Spruce to William Hooker at Kew Gardens in London, written in 1855[20]:

"You ask me about Wallace's Palms. When he came down the Rio Negro in Sept. 1851 he showed me a few figures of palms. I pointed out to him which seemed to be new, & encouraged him to go on; I also proposed that we should work them up together, I taking the literary part and he the

pictorial, which he declined. As I had also met with some of his palms and had my names for them, this caused me to relax in my study of the tribe, seeing myself likely to be forestalled in the results of my labors."

When this letter was written Wallace had already returned to England and published his classic account: *Palm trees of the Amazon and their uses.*[21] In his 1855 letter, Spruce (who must have received a copy of this book) is very critical of the illustrations and descriptions:

"The most striking fault of nearly all the figs of the larger species is that the stem is much too thick compared with the length of the fronds, and that the latter has only half as many pinnae as they ought to have. – The descriptions are worse than nothing, in many cases not mentioning a single circumstance that a botanist would desire to know; but the accounts of the uses are good."

Wallace can be forgiven for poor drawings and lack of botanical detail. On his way home to England he was shipwrecked and he lost all his specimens except for a box containing, amongst other things, his palm sketches.[22] He wrote his book from memory when he got back to England. The drawings, of which Spruce was so critical, were drawn on stone by Walter Fitch, from Wallace's sketches. Fitch added "a few artistic touches".[23] In his book, Wallace described 13 new species, five of which are currently accepted.[24]

Spruce was disappointed when Wallace declined his offer of collaboration, and, as he explained in his 1855 letter to Hooker, he would subsequently relax his efforts in palm collecting. However, it is probable that Spruce actually became more interested in palms because of the meeting with Wallace, possibly piqued by the younger, non-botanist's success on the Rio Negro. Up to that time Spruce had collected relatively few specimens of palms, even though he had been in the Amazon for over two years. He did, however, begin to make more collections in Manaus. In a letter to William Hooker dated April 18, 1851, he wrote: "*The Palms I am much interested in; they are far more numerous than at Santarem, and I believe include several undescribed species.*[25] After the meeting with Wallace, on his subsequent trip up the Rio Negro, Spruce collected at least 32 specimens, more than in any other locality. Spruce's purpose in the Amazon was to collect novelties, and after he saw Wallace's material from the Rio Negro he redoubled his efforts to collect palms, presumably also hoping to find many new species. In a letter written from Manaus to John Smith, the Curator of the Kew Gardens, dated September 24, 1851, approximately one week after the meeting with Wallace, Spruce wrote:

"I trouble you with a letter to ask you to compare the specimens of Palms I have sent to your museum with the Plates, etc., in Martius's great work and give me your opinion on them. I can find no one who will talk to me about Palms, and I am now coming among some that are exceedingly interesting. It is true that they are extremely difficult to collect and preserve. ... Higher up the Rio Negro I am certain to find

abundance of new palms. Mr. Wallace has just come down from the frontier and brought with him sketches of several palms, of which I have no doubt many are quite new.

... I am now describing completely every palm I find, and I hope to sketch the greater part of them, so that, with the aid of the specimens I send to England, I hope some day to be able to work them up."[26]

After Spruce returned to Manaus from the Rio Negro, in December 1854, he continued up the Amazon into Peru. Here he spent almost two years, mostly at Tarapoto, in the extreme western Amazon basin, near the foothills of the Andes. By this time his interest in palms seems to have waned, and he collected only about five specimens. On his subsequent ascent and descent of the Andes he collected even fewer specimens. This is interesting, because the western Amazon, especially the Tarapoto region, and the eastern Andean foothills of Peru and Ecuador are the richest areas for palms in the neotropics, and it is curious that Spruce did not collect more here, especially since he was one of the first botanical collectors in the region.

Spruce's contribution

Spruce did work up his palm collections on his return to England, and the result was an important contribution to palm taxonomy: *Palmae Amazonicae, sive Enumeratio Palmarum in itinere suo per regiones Americae aequatoriales lectarum.*[27] Although Spruce described many new species in this work, few of them are currently accepted. Spruce's contribution, however, is much greater than is suggested by the number of new species (in which endeavour he was certainly overshadowed by Martius and to a lesser extent by Wallace). Spruce is important for other reasons: firstly, because of his detailed observations and descriptions of the flowers and fruits of palms; for example, his observations on the structure of the fruits of *Geonoma* are apparently still valid although completely overlooked by recent monographers. His observations on the flowers and fruits of *Leopoldinia*, particularly his acute observation, still uninvestigated, on the similarity of fibres in the leaf sheath and mesocarp, are excellent, and he also made good ethnobotanical observations on the economically important *L. piassaba* of the upper Rio Negro.[28] His work on the flower and fruit structure, and relationships, of *Phytelephas* and *Wettinia* is also excellent, including new species of both genera.[29]

Secondly, he is important for his essay on the geographic distribution of South American palms. There is much of current interest in this essay, for example his recognition of the "granite region". It is only recently that the existence of a lowland floristic province on the Guayana Shield has been recognized.[30]

Finally, Spruce is important because he was a good botanist with great dedication and insight. Consider his discussion of the origin of the cultivated peach palm, *Bactris gasipaes* Kunth. He wrote:

"Although I am compelled to leave the native country of the Peach Palm doubtful, I quite expect the wild plant will still be met with in some

unexplored recess of the Oriental Andes, perhaps with fruit so much smaller and drier than what it has become by long cultivation as to be not easily recognizable."[31]

More than one hundred years since Spruce made that prediction, Bernal and Henderson (in prep.) have shown that the wild ancestor of *Bactris gasipaes* indeed comes from the Andes and has fruits smaller and drier than the cultivated form.

ACKNOWLEDGEMENTS

I thank Professor Mark Seaward for inviting me to attend the Richard Spruce Symposium. The manuscript was reviewed by Dr. Michael Balick.

NOTES

(1)   R.Spruce, 'Palmae amazonicae, sive Enumeratio Palmarum in itinere suo per regiones Americae aequatoriales lectarum', *Journal of the Linnean Society, Botany* (1869), **11**, 65-183.

(2)   A. de Candolle, 'Sketch on the life and writings of M. de Martius, Secretary to the Bavarian Academy of Science', *Hooker's Journal of Botany and Kew Garden Miscellany* (1857), **9**, 6-10.

(3)   J.B. von Spix and C.V.P. von Martius, *Reise in Brasilien in den Jahren 1817 bis 1820*, Munich, 1823-1831, vol. 3.

(4)   C.V.P. von Martius, *Historia naturalis palmarum, vol. 1. De palmarus generatim*, Leipzig, 1831-53; C.V.P. von Martius, *Historia naturalis palmarum, vol. 2. Genera et species*, Leipzig, 1823-1837; C.V.P. von Martius, *Historia naturalis palmarum, vol. 3. Expositio systematica*', Leipzig, 1837-1853.

(5)   A. Henderson, *The palms of the Amazon*, New York, 1995.

(6)   B. Dahlgren, 'Travels of Ruíz, Pavón, and Dombey in Peru and Chile (1777-1788)', *Publications of the Field Museum of Natural History, Botanical Series* (1940), **21**, 1-372.

(7)   F.W.H.A. von Humboldt, *Aspects of nature* (transl. Mrs Sabine), Philadelphia, 1849. [Originally published under the title *Ansichten der Natur*, Stuttgart, 1808.]

(8)   F.W.H.A. von Humboldt, A. Bonpland and C. Kunth, *Nova genera et species plantarum*, Paris, 1815-1825, 7 vols.

(9)   M. Moraes, G. Galeano, R. Bernal, H. Balslev and A. Henderson, 'Tropical Andean palms (Arecaceae)', in *Biodiversity and conservation of neotropical montane forests* (eds S.P. Churchill, H. Balslev, E. Forero and J.L. Luteyn), New York Botanic Garden, in press.

(10)   R. Spruce, *Notes of a Botanist on the Amazon and Andes* (ed. A.R. Wallace, London, 1908, 2 vols.[Reprinted edition with a new foreword by R.E. Schultes, New York, 1970.]

(11)   Martius, op. cit. (4), vol. 2.

(12)   Spruce, op. cit. (1), 65.

(13)   Ibid., 66.

(14)   Spruce, op. cit. (1).

(15)   Henderson, op. cit. (5).

(16)   Spruce, op. cit. (1).

(17)   Henderson, op. cit. (5).

(18)   A.R. Wallace, *A narrative of travels on the Amazon and Rio Negro*, London, 1889.

(19)   M. Balick, 'Wallace, Spruce and palm trees of the Amazon: an historical perspective', *Botanical Museum Leaflets (Harvard University)* (1980), **28**, 263-269.

(20)   Spruce letter to Sir William Hooker, 5 Jan. 1855, in RBG Kew Archives, Directors' Correspondence Vol. 71, doc. 372.

(21)   A.R. Wallace, *Palm trees of the Amazon and their uses*, London, 1853.

(22)   A.R. Wallace, *My life, a record of events and achievements*, New York, 1905.

(23)   Ibid.

(24)   Henderson, op. cit. (5).

(25)   Spruce, op. cit. (10), 220.

(26)   Ibid., 225.

(27)   Spruce, op. cit. (1).

(28)   R. Spruce, 'On *Leopoldinia piassaba* Wallace', *Journal of the Proceedings of the Linnean Society, Botany* (1860), **4**, 58-63.

(29)   Spruce, op. cit. (1), 1859.

(30)   For example: S. Mori, 'The Guayana lowland floristic province', *Compte Rendu Sommaire des Séances de la Societé de Biogéographie* (1991) **67**, 67-75.

(31)   Spruce, op. cit. (10), 82.

FURTHER REFERENCE CONSULTED

R. Spruce, 'On five new plants from eastern Peru. I. *Wettinia illaqueans*, a new palm from the Peruvian Andes', *Journal of the Proceedings of the Linnean Society, Botany* (1859), **3**, 191-196.

17

Richard Spruce in Northern Peru: notes on the cultivation of indigenous cotton

James M. Vreeland, Jr.

Richard Spruce in Northern Peru: notes on the cultivation of indigenous cotton

James M. Vreeland, Jr.

University of Texas, Austin, USA

While conducting field research on the native cottons of Peru's north coast and eastern jungle regions, I quite accidentally encountered a copy of Spruce's apparently forgotten paper "*Notes on the valleys of Piura and Chira in Northern Peru and the cultivation of cotton therein*".[1] The text of 81 pages was written by Spruce shortly after he definitively abandoned Peru on 1 May, 1864, and concluded on 12 September of the same year. Given the distances and difficulties of international travel, Spruce must have compiled much of the tabular information and itinerary before embarking for London where the monograph was finally edited and printed in less than six months.

Appearing at the height of the American Civil War, Spruce's cotton monograph constitutes the first systematic botanical description and comparative socio-economic account of agricultural and commercial activities, and was of enormous significance to the Manchester Cotton Association. The work is a first-class compendium of climatic, botanic, agronomic and ethnographic data, crucial for reconstructing not only regional history, but also for charting the impact of British technological and mercantile expansion on the fibre-producing economy of western South America.

Spruce's powers of observation, analysis and communication are admirably demonstrated in the monograph, with its multidisciplinary focus linking four sections, "topography and mineralogy", "climate", "indigenous vegetation" and "cotton agriculture". Clearly, the broad range of topics in this work is still of interest to natural and cultural scientists today.

Yet of perhaps more significance is how Spruce's insightful, objective and quantified observations were of vital interest to the home government at that time for the production of raw fibre for England's languishing textile industry. Spruce carefully compiled and presented historic, agro-economic and qualitative data regarding cotton fibre production and export, one of the most important agricultural commodities in the British manufacturing sector. With supplies from the American South interrupted, securing new sources of long staple fibre was considered imperative to nourish the fabric industry, starved for raw materials during the so-called "cotton famine".

Although this paper is concerned mainly with Spruce's cotton monograph, several other interesting unpublished accounts of his activities in northern Peru might also be mentioned to alert those only familiar with his pioneering work in the Amazon. His residence in coastal Peru is confirmed in the official transcripts of the United States Department of State, which maintained consuls in the towns of Lima and Lambayeque and in the northern ports of Payta and Tumbez, to monitor naval operations and maritime commerce of interest to the Union government.

Fig. 1. Detail of 18th century watercolour painted by an anonymous Indian artist depicting a Muchik woman spinning blue tinted cotton in a cone lashed to a kind of tripod support of pre-historic origin. Illustration from collection of the Bishop B.J. Martínez Compañón, Biblioteca del Palacio Real, Madrid.

In Lambayeque, capital of the province of the same name, immediately south of Piura, the American consul, James [Santiago] Coke Montjoy, reported to Washington, then at war with its cotton-producing Confederate states, on the potential production and recent growth of the sector in Peru. Of particular note to Montjoy, a southerner himself and clearly familiar with the cotton plant, was the remarkable range of natural colours of the lint, unlike anything commercially produced in the southern cotton belt. In his official report, written almost simultaneously with that of Spruce, Montjoy extensively describes the cotton culture of Lambayeque, and offers to provide the Department of State with a collection of native seeds for experimental purposes.

In the Piura valley to the north, the American consul at Payta seems to have provoked an embittered confrontation with local gentry which eventually briefly entangled Spruce. The curious affair, chronicled in the official transcripts for March, 1863, includes a testimony penned by Spruce on behalf of the beleaguered consul, who had recently been lampooned in the regional press. Despite the botanist's generous written endorsement of the American consul's moral fibre, the latter was not deterred from plagiarizing large segments of the former's notes on cotton cultivation, quoted extensively in the consul's commercial report dated 8 December, 1863, and filed with the then US Secretary of State, William Seward, in Washington.[2]

Spruce's cotton monograph provided hitherto undocumented information regarding the natural and cultural environments of the northernmost portion of

Plantación de algodón del país, de un año y medio

Fig. 2. Rare early historic photograph of native cotton plantation in Piura at the turn of the last century. The tree-like plants reached fifteen feet in height and produced as much as fifty pounds of cotton each year for over a dozen years.

the aboriginal Muchik ethnic group. With language and cultural traditions distinct from the highland Quechua and other ethnic affiliations, the Muchik culture extended across most of the desert region of northern Peru, from the Santa River in the south to the Chira River valley in the north, where it attained its greatest political consolidation about 1300 A.D. In the Piura Valley Spruce compiled a list of words attributable to one of the Muchik dialects spoken in very few north-coastal communities in the mid-19th century.

Physical setting

The geography of Peru is conventionally divided into three natural regions distributed, from west to east, in longitudinal strips best characterized as western desert coast, Andean mountains and eastern humid tropical forest bordering Colombia, Brazil and Bolivia. The continental divide lies only a few dozen kilometres inland from the shores of the Pacific Ocean, but the eastern Peruvian Amazon is several thousand kilometres removed from the Atlantic. The great uplift of the Andes during the Tertiary Period formed valleys on both flanks containing deep alluvial deposits ideal for intensive cultivation, & once populated by neolithic farming civilizations. The exuberant vegetation of the eastern mountain forests had to be cleared prior to cultivation, but the arid north coast soils are extremely productive after irrigation (systems had been established by early farming societies in the second millenium B.C.).

Based on extensive travel and personal observations, Spruce notes that the coastal plains begin north of the Ecuadorian border, widening to the south across the Sechura desert of Piura, and disappearing in the Moche valley near the modern city of Trujillo. The Piura and Chira river valleys bisect the Sechura desert plain, converting barren lands into rich oases that were, and still are, the foci of extensive cotton agriculture. The northern section visited by Spruce shows a greater degree of geological complexity.[3] The major oil fields of Lobitos, Talara, Amotape and Mancor are located where Spruce[4] documented the rich brea or mineral pitch deposits, once of great economic importance to northern Peruvian traditional societies. On the cliffs north of Payta, Spruce located important mineral, or fullers', earth deposits used to remove grease spots from articles of cotton clothing.

The comparative perspective

A perusal of Spruce's meteorological data gives a fairly complete sequence of dates upon which to reconstruct the itinerary of the botanist as he travelled both up and across the two major river valley oases (Table 1). The series begins on 9 January, 1863, and ends on 29 April, 1864, just prior to his definitive departure from Peru on 1 May of the same year. His previous residence in the province of Chanduy, coastal Ecuador, between January and August, 1862, gave Spruce an extraordinary holistic perspective of the broad environmental, and to some degree cultural, similarities of southern Ecuador and northern Peru. Spruce noted that despite the severe aridity of the region, native flora and fauna was rich and diverse, aspects often overlooked by other travellers and naturalists of

Table 1. Reconstruction of Richard Spruce's itinerary in northern Peru[7]

Piura	9 January to 8 October, 1863
Payta	17 to 22 October, 1863
Amotape	1 November to 20 December, 1863
Monte Abierto	26 December, 1863 to 25 January, 1864
Amotape	1 February to 19 April, 1864
Payta	27 to 29 April, 1864

the period. George E. Squier,[5] the American vice-consul general appointed by President Lincoln to Peru in 1862, could not hide his disappointment at the apparent monotony of the north coastal landscape when he photographed pre-hispanic ruins there in 1864.

A true scientist, Spruce employed an acute comparative perspective, focusing on the similarities between portions of the African and Peruvian desert landscapes. The wadis of the former resembled in structure and hydrology the quebradas of the latter, dry furrows in summer that become engorged and raging streams following the winter rains in the adjacent mountains. The desert environment near the mouth of the Chira river seemed to Spruce[6] comparable to that of the Indus. Both riparian regimes lacked seasonal inundation, and water had to be elevated mechanically by pumps, then channelled through irrigation networks to the surrounding plains.

Climate and "El Niño"

Spruce made climatic observations of greatest contemporary interest, documenting the general pattern of interannual rainfall and flooding of the coast, the anomalous phenomenon known as "El Niño". In "normal" years, the seasonal flooding, or "avenidas", of the Piura river was charged by precipitation falling between January and March in the adjacent Andea. These years are interrupted by years of extreme dryness and, less frequently, by epochs of extremely heavy rainfall directly on the desert coast. These "El Niño" events, according to Spruce,[8] appear to recur approximately every ten years, a pattern confirmed independently,[9] comparing similar historical sequences for the Piura region compiled from prehispanic to modern times. The major pluvial events noted by Spruce occurred in 1791, 1804, 1814, 1828, 1845, 1854 and 1864, the final year of his residence there in Peru.

Of vital importance to the indigenous population was their ability, as Spruce[10] astutely noted, to predict in advance those years of heavy rain, in order to adjust the agricultural cycle; for instance, in Amotape where he resided in November, 1863, the peasant farmers predicted heavy rainfall in the ensuing months on the basis of the strong northing winds that blew early on the mornings of the 7th and 9th. The magnitude of the irregular downpours stunned

Spruce[11] who admitted that *"even in the Amazon valley I have seen no heavier rain ..."* than in the desert of Piura in early 1864. Unfortunately Spruce did not remain in northern Peru long enough to fully document the effect of the 1864 "El Niño" event on the native flora, which had just begun to respond to the massive rains the month before his departure in April.[12]

Indigenous desert vegetation

Spruce's description of the surprisingly abundant and varied indigenous desert vegetation comprises nearly one-fifth of the monograph, which includes important morphological and ecological data for a number of species. From a cultural perspective, during "normal" years, the desert vegetation of greatest socio-economic importance documented by Spruce included a series of legumes such as the Peruvian mezquite, *Prosopis* sp., widely used by the indigenous population for food, fuel, medicinal, and construction purposes.

The aphrodisiacal properties of the bean and its effects on the local animals that consumed them were not overlooked by the ethnobotanist:

> *"One cannot walk into the potreros [pastures] without the risk of being run over by the amorous donkeys that career madly about. I am assured that a similar effect is produced on the Indian population by eating a sort of porridge which they call llupishin, made of mashed algarrobo pods and maize flour."* [13]

Spruce's visit seems to have just preceded the beginning of massive deforestation of the mezquite tree, prized for its extremely hard and durable wood and used as ties for the extensive rail tracks laid down by British and American firms in the second half of the 19th century. So hard, in fact, is the wood, that Spruce[14] marvelled at the ingenuity of farmers whose steel axes were incapable of dealing with the thick trunks. Instead, Indians resorted to an ancient practice, felling the tree with a truncheon of lighted wood placed at its base.

Cultural reconstruction

Spruce was clearly impressed by the industrious temperament of the native people, whose cultural heritage and achievement stimulated many insightful field observations. With a geologist's eye, Spruce described the stratigraphy underlying the contemporary town of Piura, built over prehispanic and colonial middens bearing valuable cultural information. In the upper levels, spanning a period of roughly 135 years since the town's founding in 1728, Spruce[15] examined remains of native cloth, footwear, food, flora and other materials preserved in the sandy subsoil matrix exposed in an eroded embankment facing the Piura river.

With a cultural historian's interpretive skill, tempered with typically Western ethnocentric conclusion, Spruce ventured a reconstruction of sartorial and alimentary customs in old Piura, based on changes in frequency of the principal remains visible in the thirty-foot high cut:

Fig. 3. Changes in native dress styles shown in this one hundred-year-old photograph of a grand-mother in traditional dress and her granddaughter in more "modern" form also fascinated Spruce who charted changes in clothing patterns in old Piura on the basis of archaeological observations. Photo H.H. Brüning, courtesy of R.P. Schaedel and the Museum für Völkerkunde, Hamburg.

"Cotton rags were everywhere common, but of a coarser quality towards the base of the deposit. Silk rags were scattered throughout, but more abundantly in the upper strata. Towards the base sandals were common, but shoes were very rare. The comparative abundance of the latter, and of the bones of oxen in the upper stata, showed the modern inhabitants of Piura both better shod and better fed than their ancestors were."[16]

The examination of extensive archaeological deposits north of Piura confirmed his supposition that massive irrigation systems built in prehispanic times allowed a greater portion of the Chira valley to be cultivated than during the mid-19th century, when it became very desertified. Now eroded, silted up and abandoned, the irrigation complex, Spruce argued, once supported large and densely nucleated aboriginal populations. His description of one such ancient canal near Mancora is worth reproducing in full:

"Not only had the water of the rivers been conveyed across ravines and along the face of steep declivities, but provision was made by canals along the base of the hills ... and ... cliffs ... for collecting the rains that fell ... and storing them up in reservoirs made by throwing strong dykes across the outlets of the ravines."[17]

Fig. 4. Spruce was a direct observer of the impact of the American and English industrial revolution which introduced labour-saving devices to north Peru, supplanting there to a large degree the laborious, hand-ginning process. This roller gin patented in Oldham, England, in 1872 is one of several hundred that still function in the Piura valley where Spruce lived between 1863 and 1864.

Agriculture and cotton cultivation

The history of cotton (*Gossypium* sp.) in Peru is perhaps one of the best documented for any non-food crop worldwide. Preserved by the arid soil matrix, the archaeological remains of primitive domesticates have been radiocarbon dated to the first half of the third century B.C. Comparable sequences reconstructed for Central America and Central Asia are nearly as old, but not so well documented, due to the much more fragmentary nature of the macro-botanical materials.[18]

Consequently, history of the genus is unique in the annals of agriculture, constituting a crop domesticated independently in at least four different continents, on opposite sides of the globe, at nearly the same evolutionary time.[19] Recent cytological analyses suggest that each domesticated species was derived from divergent progenitors by parallel and/or convergent selection for culturally desirable features, such as spinnable lint.[20]

The oldest cotton fibre remains in Peru,[21] some 4,500 years old, still retain their distinctly brown lint colour, typical of most primitive cottons in the New World. Although white lint used in yarns and fabrics clearly predominates throughout the archaeological record, brown pigmented forms are very abundant and often confused, even by experts, with cotton fibres dyed using diverse colourants. Naturally pigmented forms survive to the present day in many traditional communities of the north coast and north eastern jungle of Peru, amply documented during the past decade by the Native Cotton Project.[22]

Before arriving in Piura, Spruce had apparently already examined diverse cotton landraces in or near the probable centres of origin of each of the four modern species, the African-Asian diploids, *G. arboreum* and *G. herbaceum*, and the New World tetraploids, *G. hirsutum* and *G. barbadense*. In his travels, Spruce observed the morphology of cotton plants right across South America, including the key maritime strand ecotones through which ancient germplasm may have originally been washed, along both the Atlantic to the Pacific coasts.[22] He also documented cotton cultivation among a broad range of Indian communities exhibiting a highly diverse degree of socio-cultural complexity, and technological sophistication. Despite this marked contrast in both natural and cultural settings, spanning moist eastern tropical forest, temperate inter-Andean valleys and hot desert western plains, Spruce concluded that all ecotypes and varieties were ascribable in Peru to a single Linnean species, *G. barbadense*.

Spruce's fascination with cotton stemmed more from an economic preoccupation than from purely cultural or botanical interest: the scarcity of raw materials for the British cotton manufacturing industry might be alleviated in part through the imports of high-grade, Peruvian long-staple fibre. At the time, commercial cotton cultivation for export was undertaken on plantations using slave labour, as in the American south, until abolition in the 1850s led to the importation of Chinese coolie labour. Production was organized by large landholders, who also owned most of the cotton ginning equipment.

In the early nineteenth century, Peruvian long-staple cotton barely competed with the enormous volumes of short-staple American Upland cotton used until the blockading of southern ports by the Union navy during the Civil War interrupted exports to England. The cotton boom in coastal Peru was, however,

Fig. 5. Early historic photograph taken showing northern Peru native woman using a spinning tripod. (Compare with Figs. 1 & 6). Photo H.H. Brüning, courtesy of R.P. Schaedel, Museum für Völkerkunde, Hamburg.

Fig. 6. Recent photo of the same spinning technology employed in traditional communities of the Peruvian north coast. (Compare with Figs. 1 & 5). Photo J.M. Vreeland, Jr.

Table 2. Abbreviated record of cotton fibre exports from Piura.[25]

YEAR	QUINTALS	VALUE (PERUVIAN SOLES)
1862	3,360	198,483
1863	11,500	609,000
1864	41,455	1,245,300
1865	40,840	1,243,740
no data		
1883	37,743	692,198
1884	27,320	575,642
1885	61,250	1,341,500
1886	80,500	1,700,000
1887	49,000	818,000
1888	73,500	1,470,000
1889	63,306	1,266,112
1890	61,508	1,230,160
1891	16,953	339,060
1892	150,880	3,017,600
1893	102,290	2,045,800
1894	58,738	1,174,760
1895	44,698	893,960
1896	34,470	689,400
1897	24,574	491,480
1898	37,800	763,200

Table 3. Comparative yields of seven cotton varieties planted in northern Peru in 1863.[26]

CULTIVAR/NAME	VARIETY/ PROVENANCE	DAYS TO MATURITY	LINT PERCENTAGE	TURN-OUT
Egyptian	possibly Upland	137	32–33	3.03
Boyd's Prolific	Upland	145		
Sea Island	American	166		
Georgia	Upland	166		
New Orleans	Upland	170	32–33	3.03
Imbabura	Ecuador	174	36–37	2.78
Piura/Criollo	Peru	236	37–42	2.50

Note: turn-out is expressed as the number of units of raw cotton needed to yield one unit of clean fibre after ginning.

short-lived, following Southern Reconstruction, but reversed again briefly in the 1880s when the technical innovations in the British textile industry made possible the successful blending of Peruvian "rough" (aspero) cotton with wool.[24] The introduction of steam-powered pumps during the 1860s and the increase of land planted with cotton following the abundant El Niño rains of 1891 brought additional prosperity to the cotton sector (Table 2).

The productivity that Spruce was not able to note in his syncronic cotton monograph came from the effects of the interannual "El Niño" rainfall, occurring in 1864, which Spruce experienced, and in 1891, 1925, 1971 and 1983. Taking advantage of the periodic precipitation, peasants plant moist desert sands the year following the innunations. Bumper crops are usually harvested, as indicated for 1865, 1892, 1926, 1973 and 1984, this latter year seeing the largest harvest in the history of Piura.[27]

In contrast to other segments of the Peruvian coast, the cotton production observed in the north by Spruce was mostly undertaken by Indians, who also performed most of the ginning by hand. With the introduction of North American gins, hand-ginning disappeared entirely except for the production of homespun cloth, still made by thousands of Indian women today using pre-Columbian spinning and weaving technologies.[28] The fine quality of the Peruvian white and pigmented fibres is attested in an interesting note in which Spruce[29] makes reference to a Memorial presented to the Manchester Cotton Supply Association. Post-dated Baltimore, February, 1861, the report was written by an English planter in Peru highly recommending the north coast as a favourable locality for investment in cotton production.

It is the naturalist Spruce, however, who with an economic botanist's eye bothered to compile a convincing comparative study of yields in seven quite distinct cotton varieties and species, including Egyptian, Sea Island, Upland, Ecuadorian and native or "criollo" lines. Expressed as the percentage of lint in a unit of seedcotton, the best yields derived from the native ecotypes. Slower to mature, but no doubt heartier than the introduced cultivars, the indigenous perennial cotton of northern Peru clearly out-produced the exotic forms in an economy where labour continued to be relatively cheap (Table 3).

Spruce's[30] comments regarding biological and cultural pest control practices merit special attention by agricultural ecologists today, as "organic" cotton production is again entering fashion.[31] In particular, control for the discolouring of the fibre caused by the "cotton stainer", *Dysdercus peruvianus*, was effected simply by using soap and water. A number of similar probably prehispanic pest and disease control mechanisms continue among traditional communities in northern Peru today.[32] Staining of pigmented lint would not have posed a problem.

At the time, ginned, brown lint was sold for $14 to $15 per quintal (one hundred pounds), and was exported to France. According to Spruce,[33] the Chinese "Nankin" fibre, then commercialized in the southern United States and parts of Europe,[34] was a slightly paler brown, but equally permanent in colour-fastness. Many writers of the epoch extolled the richness of the cotton lint colours in Piura, which from time to time appeared to have fetched a higher price than white lint in the commercial market.[35]

Native cottons of Peru: demise or resuscitation?

Spruce's unpublished ethnobotanical observations also provide important comparative data on the pattern of cotton spinning and cloth production for the montane forest region of northeastern Peru. According to Alfred Russel Wallace,[36] Spruce's remarks occupy some 60 to 70 loose sheets in a stiff, paper cover inscribed with the title "Notes for a description of Tarapoto, in the Andes of Maynas or Eastern Peru".

Of central importance to the reconstruction of the history of native cotton cultivation in Latin America is Spruce's chronicle of the demise of one of ancient America's finest indigenous textile traditions, in the wake of the European industrial revolution. He eloquently characterizes the southern continent, and much of the British colonial world as well:

> "Cotton spinning is the principal industry of the women of Tarapoto. The thread is remarkably strong, and is woven by the men into coarse cloth called 'tocuyo', which used formerly to be much exported to Brazil; but latterly English and American unbleached cottons (called 'tocuyo inglez') have come hither so cheap that the native manufacture has greatly fallen off."[37]

Spruce's cotton monograph adds interesting new information to chart the progress of native cotton cultivation and cloth production in the face of poorer quality, but cheaper, imported fabrics. Curiously, in both the north coast and eastern jungle regions, the native cotton plant and craft production have survived, albeit in a much reduced scale and level of importance since the middle of the past century. On the coast, the Native Cotton Project has assembled an extensive archive of information and germplasm employed in the revival of natural cotton cultivation for speciality fibre and fabric production. In the humid eastern Andes around Tarapoto, the project has consolidated the production of several thousand hectares of indigenous cotton in three natural colours for industrial transformation.

Fascinated by the short, rather brittle reddish-brown lint, Spruce[38] remarked "the tint is rather pretty – to the Indian's eye so much so that he formerly considered it sacred, and limited its use to his priests and Incas, and to his dead". In the eastern jungle town of Tarapoto, Spruce[39] noted that the people continued to weave a strong, striped cloth for trousers and a still stouter cloth for shoes and slippers, both embellished with alternating stripes of natural white and brown cotton, some of which was then exported to other regions of Peru and to Bolivia.

Much admiring the substance and beauty of the indigenous fibre, Spruce[40] had his own clothing made from the native manufacture, which he found "very pleasant and durable wear". In the Amazon region, as on the coast, Spruce was fascinated by the several natural shades of brown cotton fibre used by the Indians to weave stripes into the coarse, homespun cloth. According to Spruce,[41]

> "the brown stripes preserve their colour after any number of washings; it is true they begin to wear rather earlier than the white, but even so it was impossible for me to get in Maynas any English or North American cottons as durable as the native fabrics".

It seems fitting to end this summary of Spruce's hallmark study of the native cottons of Peru as the naturally pigmented forms once again begin to establish themselves in the alternative, or niche, markets of Europe, Japan and the United States.[42] If Richard Spruce were alive today, he would find satisfaction in knowing that the native cotton he found so attractive and durable is not only flourishing on the Peruvian coast and jungle, but is also on the verge of a major revival. It is hoped that this brief report gives due recognition to Richard Spruce for his singular contribution to the understanding of indigenous cotton fibre and fabric production in the Andes, where the modern world learns to appreciate, once again, the achievements of traditional cultures that link Peru's resplendent pre-Columbian past with its future.

ACKNOWLEDGEMENTS

The author expresses his thanks to Dr Richard Schultes and Dr Richard P. Schaedel for their encouragement in undertaking this brief evaluation of Spruce's observations on native cotton fibre and cloth production in northern Peru, and especially to Professor Mark Seaward for his generous support and patience in this endeavour. Independent research conducted by the author on this topic has been supported by grants from the Wenner-Gren Foundation for Anthropological Research, the Inter-American Foundation, the Tinker Foundation, the Social Science Research Council, the Department of Education/Fulbright, the Organization of American States, the Institute of Latin American Studies and Department of Anthropology of the University of Texas at Austin.

NOTES

(1)   R. Spruce, *Notes on the valleys of Piura and Chira in northern Peru, and on the cultivation of cotton therein*, London, 1864.

(2)   USDS, United States Department of State, Archives, Consular Section, Peru. 1880–85.

(3)   D.A. Robinson, *Peru in four dimensions*, Lima, 1964, 158.

(4)   Spruce, op. cit. (1), 8.

(5)   E.G. Squier, *Peru: incidents of travel and exploration in the land of the Incas*, New York, 1877.

(6)   Spruce, op. cit. (1), 43.

(7)   Ibid., 16-17.

(8)   Ibid., 30.

(9)   J.M. Vreeland, Jr., 'Agricultura tradicional en el desierto de Lambayeque durante un año aluviónico', in *Actas del seminario regional de ciencia, tecnología y agresión ambiental: el fenomeno del Niño, 4-8 de junio de 1984*, Lima, Consejo Nacional de Ciencia y Tecnología, 1985, 579–624.

(10) Spruce., op. cit. (1), 22–23.

(11)   Ibid., 26.

(12)   Ibid., 42.

(13)   Ibid., 37.

(14)   Ibid., 37.

(15)   Ibid., 49.

(16)   Ibid., 49.

(17)   Ibid., 46.

(18)   K.A. Chowdhury and G.M. Buth, 'Cotton seeds from the Neolithic in Egyptian Nubia and the origin of Old World cotton', *Biological Journal of the Linnean Society* (1971), **3**, 303–312; F. Johnson and R.S. MacNeish, 'Chronometric dating', in *The prehistory of the Tehuacan valley. Vol. 4. Chronology and irrigation,* (ed. F. Johnson), Austin, Texas, 1972, 3-58.

(19)   P.A. Fryxell,*The Natural History of the Cotton Tribe (Malvaceae, Tribe Gossypieae).* College Station, Texas A. & M. Univ. Press, 1979; J.M. Vreeland, Jr., unpubl. ms.; J.F. Wendel., C.L. Brubaker, and A.E. Percival, 'Genetic diversity in *Gossypium hirsutum* and the origin of Upland cotton', *American Journal of Botany* (1992), **79**, 1291–1310.

(20)   J.B. Bird, 'Fibers and spinning procedures in the Andean area', in *The Junius B. Bird pre-Columbian Textile Conference* (eds A.P. Rowe, E.P. Benson and A.L. Schaffer) Washington, D.C., 1979, 13–17; J.F. Wendel, 'Cotton *Gossypium* (Malvaceae)', [in press]

(21)   S.G. Stephens,'A reexamination of the cotton remains from Huaca Prieta, north coastal Peru', *American Antiquity* (1975) **40**, 406–419.

(22)   J.M. Vreeland, Jr., 'Algodón del país, un cultivo milenario olvidado', *Boletín de la Sociedad Geografica de Lima,* **97**, 1978; J.M. Vreeland, Jr., 'Coloured cotton, return of the native', *IDRC Reports,* (1981), **10(2)**, 4–5; . J.M. Vreeland, Jr., 'The ethnoarchaeology of ancient Peruvian cotton crafts', *Archaeology Magazine* (1982) **35(3)**, 64–66; Vreeland, op. cit. (9); J.M. Vreeland, Jr., 'Ancient alternatives to Peru's commercial cotton pesticide crisis', *Pesticide Campaigner* (1992) **2(2)**, 1, 6; J.M. Vreeland, Jr., 'Naturally colored and organically grown cottons: anthropological and historical perspectives', *Proceedings, 1993 Annual Beltwide Cotton Conferences, Joint Session Economics/Textile Processing* Washington, D.C., 1993.

(23)   Fryxell, op. cit. (19).

(24)   R. Thorp and G. Bertram, *Peru 1890–1977 Growth and Policy in an Open Economy.* London, 1978.

(25)   F. Moreno, *Las Irrigaciones de la Costa.* Lima, 1900, 99–100.

(26)   Spruce, op. cit. (1), 57ff; for comparative historical information on American Upland varieties, see T. Affleck, 'The early days of cotton growing in the south-west', *De Bow's Southern and Western Review* (1851), **10** (orig. series), 668–669, and B.L.C. Wailes, *Report on the agriculture and geology of Mississippi.* E. Barksdale, 1854.

(27)   Anon., 'Notable baja en producción y productividad del algodonero', *El Tiempo* (1992), 25 julio, 9.

(28) J.M. Vreeland, Jr., 'Cotton spinning and processing on the Peruvian north coast. In *Junius B. Bird Conference on Andean Textiles* (ed. A.P. Rowe), Washington, D.C., 1986, 363–383.

(29) Spruce, op. cit. (1), 53.

(30) Ibid., 63.

(31) J.K. Apodaca, 'Market potential of organically grown cotton as a niche crop', Paper presented at Beltwide Cotton Conference, Nashville. 1992.

(32) J.M. Vreeland, Jr., 'Una perspectiva antropológica de la paleotecnología en el desarrollo agrário del norte del Perú', *América Indígena* (1986) **46(2)**, 275–318; J.M. Vreeland, Jr., 'Ancient alternatives to Peru's commercial cotton pesticide crisis', *Pesticide Campaigner* (1992), **2(2)**, 1, 6.

(33) Spruce, op. cit. (1), 48.

(34) J.M. Vreeland, Jr., 'Naturally colored and organically grown cottons: anthropological and historical perspectives', *Proceedings, 1993 Annual Beltwide Cotton Conferences, Joint Session Economics/Textile Processing*, Washington, D.C., 1993.

(35) Moreno, op. cit. (25), 88, 91.

(36) R. Spruce, *Notes of a botanist on the Amazon and Andes* (ed. A.R. Wallace), London, 1908, vol. 2, 83.

(37) Ibid., vol. 2, 82.

(38) Spruce, op. cit. (1), 48.

(39) Ibid., 48.

(40) Ibid., 48.

(41) Ibid., 48.

(42) Vreeland, 1993, op. cit. (22).

18

Richard Spruce's pioneering work on tree architecture

Santiago Madriñan

Richard Spruce's pioneering work on tree architecture

Santiago Madriñan

The Harvard University Herbaria, 22 Divinity Avenue,
Cambridge, Massachusetts, USA

Background

Spruce's voyage to the Amazon River

Even before Spruce embarked on his journey to South America in 1849, there was already great expectation among botanists of the 'joyfull newes' that this naturalist abroad would amass. For Spruce was not just another traveller to little-known corners of the world where, without doubt, any object of nature gathered would certainly be of great value to the scientific collections of the time. In his earlier British and European botanical research, Spruce had already shown great expertise and work of high quality. As he prepared for his new journey, the following announcement appeared in *Hooker's Journal of Botany*:

> *"Great are the inducements for a botanist to visit these [Amazonian] shores, and, influenced by these motives, Mr. Spruce proposes to set out on a voyage thither early in the ensuing spring. His object is partly for his own gratification and information, and the furtherance of the cause of Natural History; and partly that others may share in his collections, either by subscription paid in advance, to be repaid, according to priority of subscription, in specimens, at the rate of £2. the 100 species, delivered in London, or by purchasing sets at the same rate after the plants shall have reached England. Mr. Spruce also desires to supply cultivators with seeds and living plants, according to terms hereafter to be agreed upon. We believe few Naturalists have left England better qualified for the task he has undertaken, or better calculated to give satisfaction to the purchasers, than Mr. Spruce, whether in regard to the selection or the preservation of his specimens. His beautiful published collections of Plants made in the Pyrenees ... are a pledge of what may be expected of him from the regions of South America."* [1]

Spruce counted on the encouragement and support of influential botanists, including Sir William J. Hooker and George Bentham. As Director of the Royal Botanic Gardens at Kew, Hooker was much concerned with the development of economic botany and was keenly interested in Spruce's ethnobotanical observations, and in the artifacts sent for Kew's Museum of Economic Botany. Hooker published many excerpts from Spruce's letters to him, allowing botanists to follow carefully Spruce's itinerary and to be on the lookout for the latest arrivals of his collections. The publication of personal letters in journals was a common practice at the time, but was a custom abhorred by Spruce, for too often the extracts included material not intended for publication.[2] Later, Hooker played a central role in Spruce's commission from the India Office to secure seeds for the introduction of the *Cinchona* bark tree into the British Colonies of the East.[3]

Spruce sent his specimens to Kew, and on their arrival George Bentham identified all of the flowering plants and arranged and distributed the sets. Thus, Bentham had the pleasure of looking at the collections as a whole, and also of publishing the numerous new species found in them. Spruce had already given names to many of these new species, or had alluded in his notes to the novel nature of the plant. Bentham himself, in publishing them, considered the species that bore Spruce's name,[4] even though today they may be cited as "Spruce *ex* Bentham". Diligently, Bentham prepared catalogues of the plants he had received, and sent copies to Spruce. The latter apparently organized his knowledge of South American plants around Bentham's generic verifications:

> "*You mention having distributed the remainder of the Tarapoto plants, but you do not (as usual) send me the names. I trust you will do so when you next write, as nothing interests me more than to receive one of your catalogues, for although I soon forget the names I do not forget your ideas of generic affinities and distribution which they indicate.*"[5]

In his correspondence with Bentham, Spruce went into great detail on the subjects he knew would interest Bentham. On his part, Bentham related to Spruce the fate of the collections, sent detailed annual statements of Spruce's financial accounts, and reported on specific favours Spruce had requested of him, amongst which were the purchase and shipping of published works. In return for all this, Bentham received the first set of Spruce's collection.

Often Spruce entrusted his manuscripts for public presentation and publication to Bentham. This paper deals specifically with an article on the branching of trees that Spruce sent to Bentham in 1859, but first I will present as background some of the early nineteenth-century ideas on plant habit.

Humboldt and the physiognomy of plants

By the beginning of the nineteenth century, a number of botanists felt that differences in the habit of plants were of noteworthy importance. Even Linnaeus had observed "*habitus occultae consulendus est*"[6] in distinguishing between genera, although actually vegetative characters were of little importance at this level. However, in the general textbooks of the mid-nineteenth century, habit was not clearly defined. Bentham, in his *Outlines of Botany*,[7] defines it as: "The *habit* of a plant, of a species, of a genus, etc., consists of such general characters as strike the eye at first sight, such as size, colour, ramification, arrangement of the leaves, inflorescence, etc., and are chiefly derived from organs of vegetation". Despite this, botanists like Elias Magnus Fries distinguished between habit and appearance: variations in habit were connected with differences in the inner workings of the organism, and were important; variations in the appearance were trivial.[8] However, it is not clear how one could distinguish between factors relating to a plant's habit, and those relating to its appearance. Authors like J.D. Hooker were inclined to dismiss differences in habit as having little or no importance in distinguishing species.[9]

Humboldt was one of the first scientists to be concerned with the general form of plants. It is well known that he was a pioneer in the field of plant geography, further developed by Robert Brown, the de Candolles and Grisebach, among others. Humboldt also attempted a classification of vegetation based on its physiognomy, which he defines as:

> *"... whatever possesses "mass", such as shape, position and arrangement of leaves, their size ... the parts which constitute the axes (i.e. the stems and branches) ... As, then, the axes and appendicular organs predominate by their volume and mass, they determine and strengthen the impression which we receive".*

Yet the lack of technical descriptive terminology in the early nineteenth-century posed a serious barrier:

> *"Unable to depict fully according to its present features the physiognomy of our planet in this later age, I will only venture to attempt to indicate the characters which principally distinguish those vegetable groups which appear to me to be most strongly marked by physiognomic differences. However favoured by the richness and flexibility of our native language, it is still an arduous and hazardous undertaking when we attempt to trace in words that which belongs rather to the imitative art of the painter".*

Thus, Humboldt concentrated on identifying the *habitus* or *facies* of plants or of the landscape where a particular "vegetable form" predominated. Humboldt initially characterized 19 vegetable forms, including the Palms, Mimosas, Orchids, and Lianes.

Perhaps as a consequence of their inability to describe the construction and growth process of the plant, together with a stronger emphasis on holistic views of nature, Humboldt and his followers concentrated more on the physiognomic aspects of the vegetation as a whole, rather than of individual plants. This in turn led to the intuition of the Clemensian school of vegetation analysis or what may today be called Plant Ecology or Phytosociology.

Yet apparently all that what was needed to break down 'habit' into its component characters was a keen eye and a careful method of observation, coupled with the capacity to study plants in the field, particularly in the humid tropics where the most diverse array of growth forms occurs. But many botanists have visited the tropics without paying attention to the growth forms of plants, both before and after Spruce.

Spruce as collector and observer

Spruce's ample work with the mosses and hepatics while in England provided him with the experience of noting even the most minute characters. In a letter to his colleague M.B. Slater dated October 1851, Spruce writes:

> *"... [M]y Muscological studies have been of great use to me in giving me the habits of accurate and patient analysis, and after dissecting the peristomes, etc., of mosses, I find most dissections of Phanerogamia comparatively easy."*[10]

He made meticulous observations on the flowering plants he collected. His detailed descriptions included several aspects of the general physiognomy of the plant, branching, phyllotaxy, colours, textures, and the physical relationships of the parts of the flower before these were destroyed by pressing.

A transcription of a representative entry from his field books is presented below (the species is actually *Theobroma speciosum* Willd. ex Spreng):

"1737. <u>Theobroma</u> quinquinervia <u>Bern</u>. Byttner[i]æ R[io]. Janauarí. Tree, 20 f[lee]t, straight with a whorl of br[anche]s at summit twice or thrice dichot[omous]. Trunk almost completely clad with fl[ower]s, wh[ich] have fine scent of bruised lime or orange-leaves. Sepals valvate. A [illegible] fimbr[iate] disk betw[een] cal[yx] & cor[olla]. Pet[al]s 5, claws obov[ate] cuccull-saccate, pink & w[hite], limb orbic[ular] dull crimson veined, veins darker. Fil[ament]s 5, each bearing 5 or 6 cells concealed in sac of petal-claw. In centre 5 erect fleshy pub[erulan]t bl[ack]k-red tubul[ar] acum[inate] processes sterile stamens each with cuneate gland at base within. Ov[a]r[ly] w[hite] pub[erulan]t glob[bose] 5 L[i]n[es]. Style white 5 striate. - L[eave]s distich[ous] 13 1/2 x 5 1/2 obl[ong]. ineq[ual] at base abruptly apic[ulate] min[utel]y mucr[onate], ben[eath] with very close pruin[ose] w[hite] pubesc[ense]. Pet[iole]s horizontal 2–3–[?] with swollen artic[ulatio]n above middle. Guiana (Sagot)." [11]

However, as was common practice at the time, these notes were not included on the labels associated with the distributed specimens. It was only those botanists who travelled widely, and visited herbaria throughout Europe, who could examine original collectors' field books. Mez[12] extensively cites Spruce's notes in his species descriptions, particularly when referring to field characters.

Systematists of the time worked with a very limited amount of information ancillary to the herbarium specimen, covering only a few basic features of the plant. Bentham's advice to Spruce regarding note taking was:

"... the essential is to continue to put labels indicating the locality the stature the colour of the flower etc. and a number merely to serve as a means of my communicating to you the names." [13]

Spruce's notes were often made under extremely difficult conditions, but were so detailed that they could accurately evoke each collection. In a reply to Bentham,[14] Spruce writes:

"... I may refer to a remark in your memoir on Dalbergieæ that you had found intrinsic evidence in some collections of the labels having been put to the plants sometime after the date of collection. Nearly the whole of my labels have been put to the plants <u>after</u> being dried and packed; except to a very few gathered under circumstances where it was difficult to make any formal notes. My mode of working is this. When I bring home freshly-gathered plants, I make notes on them in books prepared for the purpose, and add No.'s. If any plant seems strange to me, I keep flowers &c. in water to await a spare interval when I can analyze them microscopically. So soon as the plants are dried I pack them into other paper and add the labels from my notes. As it often happens that, at each packing, I have not two plants of even the same N[atural]. Order, the risk of transposition is very small. Indeed so completely does the reading of my notes recall the features of the plants, that I feel sure if I were shown the whole of my plants classified in your herbarium, and on blank paper, I could, from consulting my notes, put to them the proper No's and localities without making perhaps a single mistake. - As to positive errors of observation I am as liable as any other mortal. - I would wish to speak with all modesty on that head; and working often in boats, or in dismal huts where a squall would suddenly enter

the open doorway & disperse both specimens & labels, there must occasionally have been some transposition of both in gathering them up again. This risk of the blowing away or dropping out of labels, was in fact what made me give up putting labels to the plants as they were drying."[15]

Characteristic of Spruce's work is the great attention he gave to the accurate description of the plants he was collecting. In his letters, he often showed concern whether he was using descriptive terminology in the right sense:

"In my remarks on the plants I have collected, I have endeavored to distinguish between "volubiles" and "scandentis" (as Endlicher seems to do), applying the latter term to such plants as elevate themselves by means of tendrils, hooked spinous stipules, & the like. In my early collections I may have called a few plants <u>twining</u> which are really only <u>sarmentose</u>. You seem to use the term "scandens" generally, which is perhaps better, as you are thus irresponsible for the precise way in which a plant may get up in the world."[16]

Indeed, Spruce's observations were so full of detail that they enabled him to venture into an area of plant biology not defined at the time, but which today we call the discipline of "Tree Architecture".

Spruce's observations on branching patterns of trees

There are many examples of Spruce's manuscripts of observations on branching patterns of the plants he collected. In referring to the forms of trunks of trees in the equatorial forests, Spruce remarks:

"Were I to unite all my observations on this head, I should be led on to write a complete treatise on the Physiognomy of Plants, which is by no means my intention here."[17]

However, it was not until late in his journey, while at Ambato in the Ecuadorean Andes, that he put together some of his notes on this subject. He prepared an essay on branching, about which he wrote to Bentham:

"Along with this letter I send you a memoir on "The Mode of branching of some Amazon trees", which I shall feel much obliged if you will read it to the Linn. Soc., and have printed in in their Journal, if thought worthy. The materials for it have been long by me, and several additional observations of living trees should have been made in order to render it more complete and systematic; but as it is hardly likely I shall ever be able to make the observations needed, I prefer sending you the deductions I have been able to make from the materials already in my hands; & I think you will find them rather interesting. If you can suggest a better title I shall be thankful; but let it by all means be a modest one, as befits the rudimentary character of the memoir."[18]

At the meeting of the Linnean Society of February 2nd, 1860, the paper was read by Bentham under its original title, and the paper published in Volume 5 of its *Proceedings*.[19]

The following extracts from this memoir highlight the links between Spruce's observations and current ideas on tree architecture.

Spruce, upon reaching the forests of the Amazon, recognized some examples of Humboldt's vegetable forms:

> "I endeavoured to trace out the species composing the forest; but, with the exception of the palms, [and] of the trees with bipinnate foliage [= Humboldt's mimosas] ... there was such an intermingling of forms that I in vain attempted to separate them."[20]

He noticed some regularity in the general outlines of the various plants he was describing, and recognized two main forms, paraboloidal or pyramidal, and obconical or obpyramidal. However, he was more concerned with the modifications in structure that would produce these shapes. He was interested in breaking down the vague term 'habit' into its components, and then in describing these components individually:

> "... what we vaguely term "habit" must depend on the disposition, form, and colour of the leaves and branches and is therefore capable of definition ..."[21]

Though Spruce may have initially talked about plants as Humboldt did (see above), he took a further step, analyzing and describing (defining) their differences. In his memoir, Spruce clearly documented examples of the different modes of branching, distinguishing variables such as phyllotaxy and branch arrangement, and indeterminate and determinate shoots. These variables constitute a contribution to what is nowadays called the "Elements of Tree Architecture.".[22]

> "[I] was forcibly struck by the paraboloidal form of the nutmeg-trees ... A closer examination showed this outline to depend on the regularly 5-nate branches, extending horizontally, pinnately branched in the same plane, the lowest or oldest branches being the longest, then gradually diminishing in length to the apex of the tree."[23]

This pattern, which Spruce calls "verticillate or whorled branching" is a clear description of what today is called Massart's model, with orthotropic main axis, rhythmic growth, verticillate branching and plagiotropic secondary axes.[24] The examples from which Spruce derived this mode of branching were all from the Myristicaceae, and he also used the name "myristicoid habit" for this growth form. In fact the Myristicaceae have such a striking way of branching that this is a very useful vegetative character in recognizing trees in the new world tropics[25] (but see below for a distinction between true "myristicaceous branching" and that common in Annonaceae). Indian communities along the Río Negro use the apical portion of the stem of myristicaceous saplings as whisks. Spruce may have noticed them in use by his Indian guides, and pondered on their branching pattern.

Warburg,[26] in his monograph on the family, consulted both Spruce's collections and his 1861 paper on the modes of branching, and he cites Spruce's work extensively in his introduction and elsewhere.

In his memoir Spruce went on to say:

> *"What I saw in Myristic[e]æ caused me to pay more attention to the mode of branching in other trees ... In most Lauraceæ however, the branches are obscurely or not at all whorled, and they ascend at various angles; so that it is rare to see, among trees of this family, any approach to the symmetrical contour of the Myristic[e]æ ... The An[n]onaceae, with much affinity to the Myristiceæ, and a very similar habit, have also pinnate branches and coriaceous distichous leaves; but the branches are solitary, not whorled."*[27]

This latter mode of branching Spruce calls "solitary or alternate branching". In this case we can equate his description to Roux's and Cook's models, i.e. a monopodial trunk with spiral phyllotaxy, continuous growth, and plagiotropic secondary axes.[28] For the Lauraceae in particular, Rauh's model differs from previous ones by including rhythmic growth and orthotropic secondary axes.[29]

The monographers for these two families, (Mez for the Lauraceae,[30] and Fries for the Annonaceae,[31]) both cited Spruce's memoir, specimens, and notes.

Spruce believed there to be a degree of regularity in branching patterns as well as a correspondence of similar modes of branching in related taxa:

> *It is unnecessary to call attention to the fact that, where the plants are similar in other points of their structure, some correspondence in their mode of branching will be found to exist, and that the differences, where there are any, have ascertainable limits in every genus and order."*[32]

Of one particular case he remarks:

> *"It is not often that we find examples of solitary and verticillate branching in the same order; and I know of but one instance of the two modes coexisting in the same genus, namely in Diospyros ..."*[33]

Here branching, together with seemingly correlated floral characters, led Spruce to consider the utilization of branching characters in generic delimitation:

> *"The differences in habit and character of the two groups are so decided, that, should the few polyandrous Diospyri found in other parts of the world possess the same verticillate ramification as the Amazon species, I should be disposed to place them in a genus distinct from Diospyros."*[34]

On this point, however, Spruce may have been misled by the limited geographical scope of his examples, for most of the species of *Diospyros* studied since then correspond to Massart's model (Spruce's "verticillate" mode). Hiern, in his monograph on the Ebenaceae, remarks:

> *"According to Mr. Spruce, his D. longifolia has branches arranged in whorls of five (very rarely three or four), while in D. paralea the branches are alternate. The branches however in D. paralea are sometimes verticillate."*[35]

Thus, according to Hiern, Spruce was mistaken; the two different kinds of branching did not even afford characters of specific value.

Spruce generally did not complement his specimen notes with drawings. Nevertheless, one particular case where a geographical representation is added to the margin of an entry in Spruce's field books is reproduced here (Figure 1).

This mode of branching is also discussed at length in his memoir and is referred to as "verticillato-proliferous ramification". Apart from *Cordia*, Spruce applies this model to Monimiaceae (in the Neotropical members of this family, all of which have decussate phyllotaxis, an approximation to this mode of branching is produced only as a reiterate response to damage of the apical meristem [pers. obs.]), *Theobroma* (see collection label 1737 transcribed above), and *Mabea*. Here the main trunk consists of a sympodium formed by subsequent orthotropic branches produced below a terminal fascicle, broadly equivalent to either Provost's or Nozeran's models.[36]

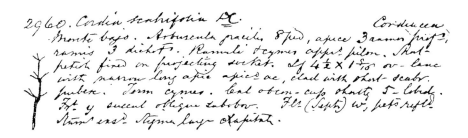

Fig. 1. Entry in Spruce's notebook of a specimen from San Carlos, April 1853 (Spruce MS: v. 2).

We could continue listing Spruce's examples and descriptions and find that each one corresponds to later architectural models such as Aubreville's (*Bucida*, *Terminalia*[37]), Corner's (*Remijia*, *Caricæ*[38]), Trolls's (*Guatteriæ*[39]), Leeuwenberg's (Spruce's obconical or obpyramidal form[40]), etc. We also find Spruce describing growth processes such as reiteration[41], monopodial (indeterminate or excurrent stems[42]) and sympodial growth (proliferous stems i.e. shoots determinate by the action of terminal inflorescences[43]).

Conclusion

Spruce's work on tree branching was not developed further; it was neglected by most of his contemporaries and as a consequence overlooked in modern works. It is not cited in the works of Corner,[44] or Hallé and Oldeman,[45] who were considered respectively as initiators and synthesizers of the discipline of tree architecture.[46] Nevertheless, it is because of the synthetic approach presented by Hallé and Oldeman,[47] later popularized in the English-speaking world by B.C. Stone's translation in 1975 and the expanded work of Hallé *et al.*,[48] that we can now perceive the originality of Spruce's contribution.

As we have seen, inquiries into the field of tree branching and plant habit had already developed by the beginning of the nineteenth century. However, not until Spruce noted characters taken from the whole plant do we find a successful attempt to describe not only the external appearance of plants, but also the growth process by which the plant attains a particular physiognomy. It is unfortunate that the botanical world of the nineteenth century, mainly concerned with herbarium-based taxonomic studies, did not encourage such field observations: full data was rarely requested or recorded[49] – as Spruce himself remarked, his work was not "compete and systematic". The dominant taxonomic discourse at the time of Spruce remained largely ignorant of habit characters, and of the underlying growth processes, or dismissed them altogether, as Hiern and J.D. Hooker did, because they were not thought to be useful systematic characters. Systematists remained focussed on that part of the plant that could be preserved on a sheet of paper of about 1300 cm^2, even though architectural characters could be useful as characters for the diagnoses of taxa.[50] It is not surprising to learn that in bryological systematics, a field in which Spruce was a leading figure, and where the whole organism can be kept conveniently in herbarium cabinets, Spruce's own classification scheme of the hepatics includes branching patterns, and androecial position, as novel and highly useful characters.[51] After Spruce's pioneering observations on tree branching, it was more than a hundred years before botanists developed the discipline of tree architecture, and made use of Spruce's insights.

ACKNOWLEGEMENTS

I wish to thank A. Cuerrier, B. Liu, Prof. R.E. Schultes and Prof. P.B. Tomlinson for comments on the manuscript, and the staff at The Botany Libraries of Harvard University for help during the review of sources. A special word of thanks to Prof. P.F. Stevens for our multiple discussions on history, systematics, plant architecture, etc. and for suggesting relevant literature. His interest in the art of historical inquiry, and concern for serving history right has been most stimulating. Any disservice here is nevertheless mine. Support for this work and travel to York was obtained through the Harvard-Universidad de Los Andes (Bogotá) Fund for Graduate Studies.

NOTES

(1)   Anon., 'Botanical information: Mr. Spruce's intended voyage to the Amazon river', *Hooker's Journal of Botany and Kew Garden Miscellany* (1849), **1**, 20-21.

(2)   Spruce to Bentham, Oct. 20 1859, in: RBG Kew Spruce papers. [Correspondence]. Letters etc. and Cinchona 1845-1880 fo. 75. Extracts from Spruce's correspondence are reproduced with the permission of the Board of Trustees, Royal Botanic Gardens, Kew. Photostatic copies consulted at The Botany Libraries of Harvard University.

(3)   Foreign Office to Hooker, 14 January 1854; L.H. Brockway, *Science and colonial expansion: the role of the British Royal Botanic Gardens*, New York, 1979, 113; R.H. Drayton, *Imperial science and scientific empire: Kew Gardens and the uses of nature. 1772-1903*, unpublished D. Phil. thesis, Yale University, 1993, 341.

(4)   P.F. Stevens, 'George Bentham and the Kew Rule', in *Improving the stability of names: needs and options*, (ed. D.L. Hawksworth) [Regnum Vegetabile No. 123], Konigstein, 1991, 157-168.

(5)   Spruce to Bentham, 9 March 1861, RBG Kew, Spruce papers, {Correspondence] letters etc. and Cinchona, 1845-80, fol. 84 recto.

(6)   C. Linnaeus, *Philosophia Botanica*, Stockholm, 1751, 117.

(7)   G. Bentham, 'Introduction: Outlines of Botany' in *Flora Hongkongensis* (G. Bentham), London, 1861, xxvii. [Appearing thereafter as an introduction to various of the colonial floras.]

(8)   E.M. Fries, *Systema orbis vegetabilis: pars I. Plantae homonemeae*, Lund, 1825, 25.

(9)   J.D. Hooker and T. Thomson, *Flora Indica*, London, 1855, 32-36

(10) R. Spruce, *Notes of a botanist on the Amazon and Andes* (ed. A.R. Wallace), New York, 1970, vol.1, 255. [Reprinted edition with a new foreword by R.E Schultes.]

(11) R. Spruce, MS. RBG Kew Spruce Papers, Plantae amazonicae 267-3846. 1849-1855, fol. 53 recto, nos. 1241-3850. [Photostatic copy at The Botany Libraries of Harvard University.]

(12) C. Mez, 'Lauraceae Americanae: monographice descripsit', *Jahrbuch des Koeniglichen Botanischen Gartens und des Botanischen Museums zu Berlin* (1889), 5, 1-556.

(13) Bentham to Spruce, Mar. 3 1850. In: RBG Kew Spruce papers. Letters to R. Spruce 1842-1890 d. 2.

(14) G. Bentham, 'A synopsis of the Dalbergieae, a tribe of the Leguminosae', *Journal of the Proceedings of the Linnean Society, Botany* (1860), **4 (suppl.)**, 2-3.

(15) Spruce to Bentham, March 9 1861, loc.cit.(2).

(16) Spruce to Bentham, October 20 1860, loc. cit. (2).

(17) Spruce op. cit. (10), 25-26.

(18) Spruce to Bentham, July 5,1859, loc. cit. (2).

(19) R. Spruce, 'On the mode of branching of some Amazon trees', *Journal of the Proceedings of the Linnean Society, Botany* (1861), **5**, 3-14.

(20) Ibid., 3.

(21) Ibid., 3.

(22) F.Hallé, R.A.A.Oldeman and P.B.Tomlinson, *Tropical Trees and Forests: an Architectural Analysis*, Berlin, 1978.

(23) Spruce op.cit.(19), 3.

(24) Hallé, Oldeman and Tomlinson op.cit. (22), 191.

(25) A.C.Smith, The American species of Myristicaceae', *Brittonia* (1937), **2**, 393-510; A. Gentry, *A Field Guide to the familes and genera of woody plants of northwest South America (Colombia, Ecuador, Peru) with supplementary notes on herbaceous taxa,*. Washington, DC, 1993.

(26) O.Warburg, 'Monographie der Myristicaceen', *Nova Acta Academiae Caesareae Leopoldinino-Carolinae Germanicae Naturae Curiososum* (1897), **68**, 1-680.

(27) Spruce, op. cit. (19), 4-5.

(28) Hallé, Oldeman and Tomlinson op. cit. (22), 200, 206.

(29) Ibid., 221.

(30) Mez, op. cit. (12).

(31) R.E. Fries, 'Revision der Arten einiger Anonaceen-Gattungen I-V', *Acta Horti Bergiani* (1930-1939), **10**, 1-341; **12**, 1-577.

(32) Spruce, op. cit. (19), 13.

(33) Ibid., 7.

(34) Ibid., 7-8.

(35) W.P.Hiern, A Monograph of Ebenaceae. *Transactions of the Cambridge Philosophical Society* (1873) **12**(1), 241.

(36) Hallé, Oldeman and Tomlinson, op. cit. (22), 161, 177.

(37) Spruce, op. cit. (19), 11.

(38) Ibid., 21.

(39) Ibid., 5.

(40) Ibid., 8-10.

(41) Ibid., 13.

(42) Ibid., 12.

(43) Ibid., 12.

(44) E.J.H. Corner, 'The Durian Theory or the origin of the modern tree', *Annals of Botany.* n. s.(1949),**13**, 367-414; E.J.H.Corner,'The Durian Theory extended - I', *Phytomorphology* (1953), **3**, 465-476; E.J.H.Corner, The Durian Theory extended - II. The arillate fruit and the compound leaf. *Phytomorphology* (1954), **4**, 152-165; E.J.H.Corner, 'The Durian Theory extended - III. Pachycauly and megaspermy - Conclusion', *Phytomorphology* (1954), **4**, 263-274.

(45) F. Hallé and R.A.A.Oldeman, *Essai sur l'architecture et la dynamique de croissance des arbres tropicaux*, Paris, 1970. [English translation by B.C. Stone, Kuala Lumpur,1975.]

(46) C.Edelin, 'The monopodial architecture: the case of some tree species from tropical Asia',*FRIM Research Pamphlet* (1990), **105**, 229 pp.

(47) Hallé and Oldeman op. cit. (45).

(48) Hallé, Oldeman and Tomlinson, op. cit. (22).

(49) P.F.Stevens, *The development of biological systematics: Antoine-Laurent Jussien, nature, and the natural system*, New York, 1994.

(50) Hallé, Oldeman and Tomlinson op. cit. (22).

(51) B.M. Thiers, 'Introduction and index with updated nomenclature', in reprint of 'Hepaticae of the Amazon and the Andes of Peru and Ecuador' (R. Spruce), [reprint], *Contributions from the New York Botanical Garden* (1984), **15**, i-xii.

19

Relevance of Spruce's work to conservation and management of natural resources in Amazonia

N. J. H. Smith

Relevance of Spruce's work to conservation and management of natural resources in Amazonia

N. J. H. Smith

LATEN World Bank, 1818 H St, NW,
Washington, D.C. 20433, USA

Richard Spruce has provided us with revealing glimpses of landscapes and how people were using natural resources in the Amazon during the middle of the last century. I have plumbed Spruce's writings, particularly his *Notes of a botanist on the Amazon and Andes*, with two main purposes in mind: to glean observations on crop plants and their wild populations; and to extract interesting ethnobotanical observations as pointers to possible candidates for plant domestication.

Approximately half of the increased yield of major crops in this century is attributed to harnessing of genetic resources. Further productivity gains will only be possible if we conserve and utilise plant genepools. At the same time, we should be constantly seeking new crops to help us face the challenge of tailoring our suite of cultivated plants to difficult environments, and to provide novelty in the diet of industrial and developing countries alike. Novel crops, then, could play a vital role in boosting incomes for rural inhabitants in developing areas while delighting the palates of well-fed consumers in the industrial world.

As both industrial countries and developing nations seek ways to raise crop yields without damaging the environment, scientists constantly tap genes to develop crop varieties more resistant to pests and diseases.[1] Plant breeding alone is no panacea for agricultural woes, nor a complete blanket of protection for our farmlands, but when combined with fresh agronomic approaches, such as rotation cropping and biocontrol using parasites and predators of crop pests, the manipulation of genetic resources often holds the key to upgrading farm productivity.

Plant breeders tap three main genepools when working to improve crops. In descending order of importance they are: the primary genepool, consisting of material in the same species, such as traditional varieties; related genera; and finally, plants in different families.

With more sophisticated "genetic engineering" techniques, it is now increasingly possible to insert genes from one species to another. Such skills, however, rest on a thorough understanding of the relationships between crop cousins, and Richard Spruce made a number of important discoveries in this regard, especially in connection with rubber and cinchona. He also made numerous interesting observations on so-called minor, or little known forest products that could prove useful for the future development of plantations and agroforestry.

Major plantation crops

A. Rubber

Six new species of *Hevea*, the genus that includes the most important source of natural rubber – *H. brasiliensis* – were described on the basis of Spruce's collections in Amazonia.[2] Near relatives of rubber have already been used to improve the hardiness of rubber; for example, *H. spruceana* is grafted onto rubber trees to provide a crown resistant to South American leaf blight, caused by *Microcyclus ulei*, a widespread fungus in the Amazon Basin and the most serious disease threatening plantation rubber. Should South American leaf blight ever reach Southeast Asia, where most of the world's natural rubber is currently produced, a veritable scramble for resistance genes would ensue. Spruce's pioneer work on the taxonomy and distribution of natural rubber and its near relatives would thus become even more valuable. Richard Evans Schultes, who has done more than anyone else in this century to further our understanding of rubber's genetic resources, has emphasised how all consumers of rubber products are indebted to the work of Spruce.

Few people who drive cars with radial tyres realise that they contain sizeable amounts of natural rubber. Nor are aeroplane travellers generally aware that when they touch down, their landing is cushioned by tyres made entirely of natural rubber. Few appreciate, then, the contribution of Amazonia's forests and one of its great explorers, Richard Spruce, to modern life.

B. Cinchonas

Spruce also furthered our understanding of another economically-important tree: *Cinchona*, which grows in western Amazonia, in the forests cloaking the footslopes of the Andes. The bark of several species of *Cinchona* contain quinine, and long ago indigenous peoples found that they could relieve certain fevers by chewing pieces of the bark.

Spurred on by quinine's anti-malarial properties, the British government commissioned Spruce to prospect for *Cinchona* planting material in 1857. After securing permission from Ecuadorian authorities, Spruce set out over rugged terrain to collect botanical specimens and seeds of *Cinchona*, to obtain bark samples, and to note their distribution and natural history. Spruce's body was almost spent after the rigours of nearly a decade in the backwoods of Amazonia, yet he persevered in his quest for specimens of widely-spaced *Cinchona*. Spruce's work was fundamental to the success of *Cinchona* plantations in various parts of the highland, humid tropics, especially in India, Sri Lanka, and Indonesia.

At first glance, such travails may seem a rather quaint side-story in the annals of economic botany, except that quinine is re-emerging as an important pharmaceutical. Schoolchildren in England during the 1950s and early 1960s may remember swallowing, with great reluctance, bitter doses of quinine to treat colds. Such memories may create the impression that quinine is history. But the emergence of strains of *Plasmodium falciparum*, the most virulent species of malaria parasite, resistant to a range of modern drugs, has brought quinine back into the picture. Quinine is once again a routine part of malaria therapy in various parts of the tropics, sometimes in combination with other drugs, such as Fansidar (a combination of sulfadoxine and pyrimethamine) or tetracycline. Hundreds of thousands, if not millions, of lives have been saved because of quinine over the last century or so, and part of the credit belongs to Richard Spruce.

C. Brazil nut

As forests in Amazonia continue to fall and the once extensive populations of Brazil nut (*Bertholletia excelsa*) shrink (Figure 1), several entrepreneurs have recently begun experimenting with plantations of the nutritious nut. Cattle ranchers are spearheading the attempt to grow Brazil nut on a large, commercial scale, such as near Itacoatiara in Amazonas state (Figure 2). Many small farmers throughout the Brazilian Amazon are incorporating Brazil nut into their innovative agroforestry systems, such as around Tomé-Açu in Pará and Nova Califórnia in Acre/Rondônia.

Brazil nut planting is not a new undertaking, however. Indigenous groups have long enriched various parts of the Amazon Basin with majestic Brazil nut trees, but this realisation is relatively recent. Careful readings of some of the natural historians of the last century and even earlier often reveal useful ethnobotanical findings, and Richard Spruce is no exception. Spruce[3] noted, almost in passing, that Jesuits planted Brazil nut at Tauaú along the Acará River in Pará. Jesuits were operating missions in the Brazilian Amazon until expelled by Royal edict in 1759; so the dark-robed Fathers were evidently emulating some Indian horticultural practices in the early colonial period. Ancient groves of Brazil nut can be seen around various towns and villages in Amazonia today, such as Acará on the river of the same name, not far from the site visited by Spruce in the last century. Spruce also remarks that agoutis, pacas, and monkeys eat Brazil nuts; the agoutis are known to be dispersal agents for the forest giant.

Minor products from forest and other habitats

Spruce hints that some forests and other habitats in Amazonia have probably been enriched by indigenous groups. He remarks, for example, that the forests around San Carlos on the upper Rio Negro are liberally sown with patauá palm (*Oenocarpus bataua*).[4] The superior qualities of oil from fruits of patauá palm have long been recognised, and the potential for planting the moisture-loving palm as a commercial crop is now being explored by Brazil's agricultural research system (EMBRAPA – Empresa Brasileira de Pesquisa Agropecuária) in Pará. So fine is patauá's oil that shopkeepers in Belém used to mix it in equal proportions with olive oil during the last century, and sell it to unsuspecting customers. If water-loving patauá takes off as a commercial crop, breeders and growers will be interested in a wide selection of material to test. What better place to begin the search for promising germplasm than anthropogenic forests where indigenous people may have selected for more productive forms?

Another oil-bearing palm, mucajá (*Acrocomia aculeata*), has undoubtedly been sown around homes and villages for millennia. The smooth, creamy pulp surrounding the seed is still relished as a snack food by rural peoples in Pará

Fig. 1. Forest with Brazil nut trees (*Bertholletia excelsa*) drowned by the Tucurui reservoir, Tocantins River, Pará, Brazil. In 1986, the dam began generating much-needed hydroelectric power, but has eliminated extensive populations of Brazil nut trees and other economically-valuable species. Picture taken in August 1988.

Fig. 2. A grafted Brazil nut (*Bertholletia excelsa*) plantation at Fazenda Aruanã, km 215 Manaus-Itacoatiara, Amazonas, Brazil, 15 November 1992. The 4,000 hectare Brazil nut plantation has been established on degraded pasture and the grove depicted here is six years old. Strips of forest have been left as havens for bee pollinators.

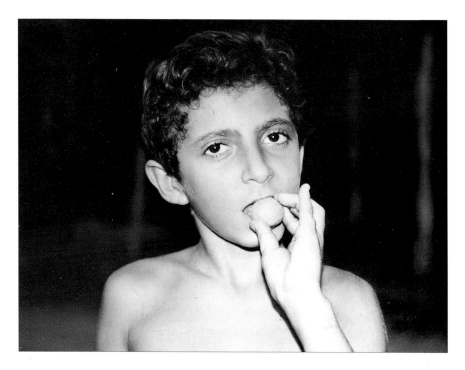

Fig. 3. Farm boy snacking on creamy fruit of mucajá (*Acrocomia aculeata*), collected from the ground under a spontaneous palm on the border of a field, Belterra, Pará, Brazil, 29 September 1992.

and is fed to small livestock, such as at Belterra, on the *planalto* south of Santarém (Figure 3). Mucajá's sweet-tasting mesocarp contains 33 percent oil;[5] little wonder that people have sought the nutritious palm and enriched their home sites with it. Interest in fully domesticating the species is growing in Brazil and in growing it on a larger scale for oil production.[6] Mucajá populations have probably built up from discarded seed, although some individuals, or their ancestors, may also have been deliberately planted. Spruce[7] speculated that the palm, also known locally as macaúba, had been planted inland from Belém, since it is only found in open habitats near dwellings.

Although Spruce was primarily concerned with describing flora in wild places, he took the trouble to point out cultural uses of plants; for example, he noted that the durable fronds of ubím palm (*Geonoma*) were used for thatching houses on the lower Amazon. Another palm used for construction purposes along the lower Amazon is bussú (*Manicaria saccifera*). Both rural and urban folk along the lower Tocantins continue to use the generous fronds of bussú to cover their houses and huts used to make manioc flour (*casas de farinha*). Spruce[8] noted that leaves of the latter palm are entire, thus forming long "tiles" when laid out on a roof, ideal for dispelling rainwater.

People in Amazonia collect plant products from a wide variety of vegetation communities, not just forest, and Spruce was clearly attuned to this wide-ranging dependence of rural people on nature. A tall, handsome grass (*Gynerium sagittatum*), which resembles a slender sugarcane, adorns some open parts of the banks of the Amazon and its side-channels. Called *flecheira* on the Amazon floodplain today, Spruce noted its Tupí name, *uíwa*, and pointed out its use to make arrow shafts. Spruce often included both the local name (mostly *lingua geral*, in his day) and the indigenous name (usually Tupí) when talking about plants and animals. Such information is helpful, and he also took pains to point out how place names have changed over time. His *Notes of a botanist on the Amazon and Andes*, lovingly put together by one of his admirers, Alfred Russel Wallace, is thus of considerable historical importance in addition to being a fountain of knowledge about plant geography and ethnobotany.

In other open habitats, fisherfolk used to collect leaves of ajarí (*Tephrosia nitida*) to stupefy fish. Spruce[9] remarked on how the leaves of the leguminous herb were crushed in water to facilitate fishing. Several other plants are collected or are grown for that purpose in Amazonia, and Spruce noted that *Tephrosia toxicaria* was cultivated for its piscicidal properties around Santarém. The practice of cultivating plants with an ability to stupefy fish is now much less common, in part because of the advent of more "efficient", but far more destructive, fishing methods, such as gillnets and the illegal use of dynamite.

Spruce's ethnobotanical notes point to some intriguing candidates for plant domestication. Along the Tapajós, women used to grate the seeds of the acapurana (*Campsiandra* sp.), a low, spreading leguminous tree, to make a flour, which served as a substitute for manioc flour when supplies of the starchy root crop were running low.[10] The genus *Campsiandra* is being revised, and it appears that several species produce fruits that are consumed in southern Venezuela and in the Brazilian Amazon (B. Stergios, pers. comm.). Fruits of acapurana are still occasionally used in folk remedies,[11] but this denizen of the banks of clear and blackwater rivers, as well as open sandy soils, is apparently little used today as a 'famine food'. Encouragingly, Venezuelan researchers at the Universidad Nacional Experimental de los Llanos Occidentales are investigating the current use and potential of protein rich *Campsiandra* flour.[12] This perennial legume, which once served as emergency rations, might prove suitable for agroforestry on floodplains, or in areas with poor soils.

Wild populations of crop plants

Wild or feral populations of crop plants sometimes contain useful genes absent in domesticated genepools. Breeders are increasingly interested in the location and status of wild or naturalised populations of crops, because they may have been lost after prolonged breeding and selection in cultivated populations, or they may not have been present in the initial material used for domestication. Spruce remarks on the presence of several wild populations of crop plants, potentially useful leads for future collecting efforts.

The eastern slopes of the Andes abound with wild relatives of succulent papaya (*Carica papaya*). Two near relatives of papaya have also been domesticated. Spruce observed chambúru (*Carica pentagona*) under cultivation up to 3,000 metres in the Ecuadorian Andes, where people ate the fruits and boiled the leaves like cabbage.[13] Other species might eventually be domesticated, or their genes incorporated into cultivated forms. In eloquent prose, Spruce expresses his respect for nature and his awe of its biodiversity when discussing the papaya family:

> *"Where Papayaceae most abound is on the wooded slopes of the Andes, both on the eastern and western sides, up to 8,000 feet elevation; and it is there that travellers and sedentary botanists may confidently expect to find not only materials for the more perfect elucidation of the species already partially known, but also many new species, which doubtless still remain hidden in the savage recesses of the oriental Andes".* [14]

Cacao (*Theobroma cacao*) is native to western Amazonia, and was taken to Central America in pre-contact times where it was developed as a crop to produce chocolate. This understorey perennial does not appear to have been an important economic plant in Amazonia prior to the arrival of Europeans. The shade-tolerant tree was mostly enrichment planted in forest near villages and camps for its sweet pulp. In the 18th century, cacao planting received a boost as an export commodity and large areas were planted along the middle and lower Amazon. By the mid-18th century, however, cacao had lost some of its former commercial importance and was in decline. Spruce[15] noted a large plantation on Careiro Island near Manaus being overrun by forest. Cacao was apparently the predominant species in floodplain forest of the Amazon south of the massive Ilha dos Tupinambás, near the present-day town of Parintins. Cacao may not be indigenous to that part of Amazonia, suggesting that some parts of the várzea forest have been cleared in the past and are therefore anthropogenic. Cacao populations in floodplain forests should be a high priority for collecting because woodland on the várzea is the most threatened forest type in Amazonia, and pockets of feral cacao may still be encountered.

The tart fruits of pitomba (*Talisia esculenta*) are a welcome treat in Pará during the rainy season. The marble-sized fruits are commonly sold in bunches in street markets in such cities as Belém and Santarém. Mostly grown as a backyard tree and occasionally in polycultural fields, pitomba is one of many regional fruits in Amazonia that warrants further study. Pitomba is apparently indigenous to western Amazonia, where it may occur in the wild.[16] Spruce[17] spotted wild, or at least naturalised, populations of pitomba on stony slopes of the Tapajós valley, in the central part of the Amazon Basin. Such populations − if they have been spared from the axe and chain saws of small farmers and ranchers – could prove useful sources of germplasm for tolerance to especially poor soils. Agroforestry systems suitable for rehabilitating degraded areas in the tropics are in high demand, and observations on the location and habitat of wild populations of crop plants are thus especially important.

Lessons for sustainable development

Spruce's vibrant writings allow us to see plant life, people, and landscapes of Amazonia in the middle of the last century. Modern scientists and policymakers may well ask what an eccentric botanist of yesteryear has to offer us today in the quest for more rational models for occupying and developing the Amazon.

Spruce's keen observations on the natural history of some plant species that may be tomorrow's key crops in Amazonia and his extensive collections serve as a valuable foundation for sustainable development in the region and other parts of the humid tropics.

A major lesson that can be extracted from Spruce's life and work was his dedication to prolonged field work and the quality of the information he recorded. An intimate knowledge of people and their resources is essential for rational development of any region.

After collation and analysis of existing literature and statistical manipulation of available data (essential ingredients of any modern research programmes) ground-truthing and uncovering primary information are still essential for a better informed picture of a region's potential for development and conservation, particularly in the humid tropics where official statistics are frequently unreliable or non-existent, and other data incomplete. Although rural populations may know a great deal about their environment, they are sometimes wrongly perceived as ignorant, with little information to offer in efforts to raise living standards in a region. The push for "desk-top" syntheses and computer model simulations can be almost irresistible as a supposed cost-saving device. The value of field work is often poorly appreciated by scientific directors and administrators of government agencies and development organizations; and it can be hard to get funding for field work because it is sometimes considered less prestigious or important than laboratory and computer-based work. But although field work is costly, the ultimate price for not understanding the natural resource base of a region and how people interact with nature before interventions are planned can be very much higher.

As I read through Spruce's epic book of travels in Amazonia, I was awed by how much time he spent in the field. He would sometimes spend two or three months waiting for a sail-boat to take him and his supplies up-river. Rather than remain house-bound, however, Spruce launched off almost daily into surrounding fields and forest to collect plants and observe the landscape. Heavy rains hardly deterred him, nor biting black flies and mosquitoes. Now there are as many as two jet flights a day from Santarém to Manaus, and the journey takes an hour. Spruce spent two months sailing from Santarém to Manaus. Modern travel is both a blessing and a bane. Jets allow us to reach our destinations so much quicker, but we miss much of the countryside on the way. May the example of Spruce's life help kindle a passion for field work in young scholars and all those committed to safeguarding and more effectively using the biodiversity of Amazonia.

ACKNOWLEDGEMENTS

I am grateful to the World Bank for supporting my participation at the Spruce meetings at the University of York in September 1993, and for preparing prints from some of my slides. The views and conclusions in this chapter are mine and do not necessarily represent those of the World Bank or any other organization.

NOTES

(1)   T. T. Chang. 'Availability of plant germplasm for use in crop improvement', in *Plant breeding in the 1990s* (eds H.T. Stalker and J.P. Murphy), Oxford, 1992, 17-35; J.G. Hawkes, *The diversity of crop plants*, Cambridge, Mass., 1983; N.J.H. Smith, J.T. Williams, D.L. Pluckett and J.P. Talbot, *Tropical forests and their crops*, Ithaca, 1992; G. Wilkes, 'Germplasm preservation: objectives and needs', in *Biotic diversity and germplasm preservation: global imperatives*, (eds L. Knutson and A.K. Stoner), Dordrecht, 1989, 13-41.

(2)   R.E. Schultes, 'Richard Spruce still lives', *Northern Gardener* (1953), **7**, 20-27, 55-61, 87-93, 121-125. [Also issued as repaginated reprint, pp. 1-27.]

(3)   R. Spruce, *Notes of a botanist on the Amazon and Andes* (ed. A.R. Wallace), London, 1908, vol. 1, 16.

(4)   Ibid., 479.

(5)   C. Pesce, *Oil palms and other oilseeds of the Amazon* (ed. and transl. D.V. Johnson), Algonac, Mich., 1985, 28.

(6)   L. Coradin and E. Lleras, 'Coleta de germplasma de macaúba: situação atual', *Newsletter: Useful plants of tropical America* (FAO/CENARGEN) (1986), **2**, 5-6.

(7)   R. Spruce, 'Palmae Amazonicae, sive Enumeratio Palmarum in itinere suo per regiones Americae aequitoriales lectarum', *Journal of the Linnean Society, Botany* (1869), **11**, 65-183.

(8)   Spruce , op. cit. (3), 59.

(9)   Ibid., 86.

(10)   Ibid., 149.

(11)   M.F. da Silva, P.L.B. Lisbôa and R.C.L. Lisbôa, *Nomes Vulgares de Plantas Amazônicas,* Manaus, 1977, 12.

(12)   B. Stergios, 'La etnobotanica del arbol "chiga" (Campsiandra, Leguminosae, Caesalpiniaceae) en la region llanera de la cuenca del medio Rio Orinoco en el Suroeste de Venezuela', *Biollania* (1993), **9**, 71-90.

(13)   J.C. Mello and R. Spruce, 'Notes on Papayaceae', *Journal of the Linnean Society, Botany* (1989), **10**, 1-15.

(14)   Ibid.

(15)   Spruce, op. cit. (3), 233.

(16)   P.B. Cavalcante, *Frutas comestíveis da Amazônia,* Belém, 1991, 193.

(17)   Spruce, op. cit. (3), 162.

20

An anthropologist's debts to Richard Spruce

G. Reichel-Dolmatoff

An anthropologist's debts to Richard Spruce

G. Reichel-Dolmatoff

University of California, Los Angeles, USA

It has been my privilege and good fortune to spend much of my life in Colombia – 55 years all together – where I have been active as a field anthropologist in many rain forest areas. Half a century ago there were very few studies of Colombia's many tribal societies. There were, of course, the early Spanish chroniclers, and some later reports written by missionaries or colonial administrators; but the truly important bibliographical sources on the Indians were books and research articles written by nineteenth-century travellers, most of them Europeans.

Colombia has been called the "Gateway to South America", because of its geographical position, and it is true that many travellers and explorers entered Colombia from the North, that is from the Caribbean, travelling up the Magdalena River to reach Santa Fé de Bogotá, the country's capital, high up in the mountainous interior. In the eighteenth century the Spanish botanist José Celestina Mutis travelled up the Magdalena, while the Austrian botanist Nikolas Joseph Jacquin collected on the lower Magdalena River. In the first years of the nineteenth century came Alexander von Humboldt and Aimé Bonpland. These pioneers were followed by a large number of naturalists. Radiating out from Bogotá, they visited the inter-Andean valleys, the cordilleras or the Pacific Coast, and a large area of the Colombian interior. Many of the books written by them in English, French, German, and Spanish have been translated and re-edited, and so have reached a wide readership.

But the Magdalena River was not the only entrance to Colombia. Many naturalists, and travellers entered what eventually became Colombian territory by way of the Amazon, a fact hardly known in Colombia today, and even if known, insufficiently appreciated. The enormous area of the Amazon/Orinoco watersheds (more than half of Colombia's territory) is almost totally underdeveloped. Most Colombians have only the vaguest idea of the geography, flora and fauna of this part of the country, and do not realise that it is peopled by thousands of Indians who continue to speak their own languages and practise many of their traditional customs.

It was in this area that the great naturalists did much of their work, among them Richard Spruce, Alfred Russell Wallace and, farther south, Henry Walter Bates. Thirty years before them, the Spix and Martius Expedition had already penetrated up the Caquetá and Putumayo rivers, deep into what is now Colombian territory. None of these men ever went up to Bogotá, and their work is therefore known only to botanists and zoologists. It is a great pity that this should be so, because the publications of these early naturalists who entered Colombia from the Amazon contain a wealth of ethnographic information and, moreover, a wealth of intellectual stimulation. Their books are most useful sources on many aspects of indigenous societies, and it is a pity that most

Colombians, including anthropologists, seem to be quite unaware of this specific rain forest literature, in which sound scientific observation is combined with personal engagement, and good writing.

In the middle of the last century the British naturalists Bates, Spruce, and Wallace, who worked in the northwest Amazon, opened up a vast field for botanical and anthropological research, which has never been described so vividly, earnestly, colourfully and humanely as by these travellers.

Spruce holds the reader's attention by the way he describes the rain forest and its inhabitants. He writes as both botanist and sensitive European, of the impact tropical nature made upon him: Spruce wrote: *"There were enormous trees, crowned with magnificent foliage, decked with fantastic parasites".*

Or, when speaking of a certain landscape on the Rio Negro, the Cerro de Cocuí, he wrote: *"... it seemed the finest object for a painter's pencil I had seen in South America. It is impossible to do justice to the scene in words".*

When I was in my twenties, reading these books for the first time, I imagined a perfumed Arcadia such as Wallace described, when living among the Indians of the Colombian Northwest Amazon, he found himself at times in what he calls: *"... A stage of excited indignation against civilised life".*

My main interest was the Indians, their belief systems, their ways of interpreting the universe, with their cosmology, their shamanistic practices. However, I realised that the key to understanding these beliefs and practices was the plants involved. This became clear to me very early on, and it was in reading Spruce and Wallace that I became fully aware of this; it was these Victorian travellers who awakened in me a lasting interest in botany and in the rain forest.

Now my knowledge of botany is limited. Nevertheless, I have always believed that any aspect of a traditional, native society must be seen in the full context of its natural environment, and that the anthropologist must learn how the Indians relate to plants and to wildlife; not only in terms of planting and harvesting, of hunting and fishing and foraging, but in terms of sharing with plants a common territory, of living together, of participating in a common source of vital energy, whatever its name, however defined or conceptualised. This relationship, it seems to me, must be understood not only in terms of a practical use of plants, of their economic or caloric value, their medicinal use, but also in terms of abstract meanings, of the symbolism of plants, of their metaphoric value. Plants – their morphology and physiology, their chemical properties, their colours and odours, flavours and textures – constitute in the Indians' minds a coded system of communication in which practically all aspects of social values and norms are linked together.

The most important plant of the Northwest Amazon from a caloric and economic point of view is bitter manioc (*Manihot esculenta*). From a spiritual point of view, I would say that the most important plants belong to the genus *Banisteriopsis*, the hallucinogenic vine first described by Spruce. Botanical and pharmacological knowledge of this genus has been greatly advanced by the studies of Professor Richard Evans Schultes and his associates, but, more than that, *Banisteriopsis* studies have come to form an important bridge between botany and anthropology, and Schultes' work on Northwest Amazonian narcotics (inspired by Spruce) is undoubtedly a major contribution to anthropology.

Drug-induced trance states are fundamental in many aspects of Indian life. Among the Indians I know, *Banisteriopsis* is not thought to be a medicinal plant, but is mainly related to spheres of cosmology, religion, philosophy, social organisation, to artistic manifestations and the world of imagery. The perception of the hallucinatory dimension depends entirely upon the local culture, upon a local consensus, and Richard Spruce was quite aware of this when he observed the use of *Banisteriopsis* among the Indians of the Orinoco plains, then among those of the Rio Negro rain forest, and those of the Ecuadorean *montaña*. Further studies of Northwest Amazonian narcotic plants are likely to make important contributions in the fields of pharmacology, neurophysiology, psychiatry, and anthropology.

The study of the shamanistic use of *Banisteriopsis* to which Spruce and his modern followers introduced me was certainly not the only research topic to which I was led by the early botanists. Spruce had been much interested in palms, a family I had previously looked on merely as a source of food, of oil, or as raw material for some native manufactures. Reading Spruce and Wallace, I began to wonder about palm symbolism, about the palm as metaphor. I began to ask myself: "Why should the Indians elaborate on the importance of certain palms while almost ignoring others? Why should there be so much talk about stilt-roots or about the bulging stems of *Ireartea ventricosa*? Or why should the Indians discuss in detail the peculiar smell of the pollen of *Socratea exorrhiza*?

I read up on palm physiology and pollination, on fruiting seasons and geographical distribution. Occasionally the writings of the botanists exasperated me because they did not mention details which to the Indians were of importance in their botanical systematics. However, eventually I was able to ask the Indians intelligent questions – 'intelligent' in terms of the Indians' understanding of things. In this way I began to enter another dimension of palm lore; not that of palms as a food resource, but of palms as a model, an icon, of palms the fruits of which were not edible, but which belonged to a large category of trees which were of extraordinary symbolic importance. I would never have made a study of the mental processes underlying this palm imagery, were it not for the Northwest Amazonian botanists and their interest in palms, Richard Spruce having been a pioneer in this field.

Spruce mentions two large forest trees, the *uacú*, as it is called in the Amazonian vernacular (*Monopteryx angustifolia*), and the *japurá* or *Erisma japura*. He traces their distribution up the Rio Negro to the Casiquiare, and it would seem that these trees particularly attracted his attention. It is true that the Indians often mention these two trees and their edible seeds, but detailed botanical descriptions seem to be scarce. The trees grow scattered here and there, and outsiders usually do not think they are of any importance.

However, to the Indians, these two trees, their fruiting season, and the foodstuffs prepared from their seeds are more important than many other forest resources, and relate to many social aspects of their lives which have symbolic expression in these trees. The high protein content of the seeds is fully recognised by the Indians and is ritually manipulated in a form similar to other protein-rich food resources. I should add here that the *uacú* tree (*Monopteryx angustifolia*) is often mentioned in Amazonian mythology as a kind of "Tree of

Life", an image which can be understood only if one knows the Indians' interpretation of the tree's morphology and physiology, together with the processing of the seeds and the beliefs connected with their consumption.

In conclusion, Spruce has offered me a glimpse of a New World and made me ask new questions; what more could one demand from a great teacher? Richard Spruce has been a trusted guide in the Northwest Amazon, and has introduced me to rain forest botany, and to many botanists, old and young, whose friendship and collaboration have been important in my life. I am grateful to the memory of that extraordinary man, Richard Spruce.

21

Richard Spruce's economic botany collections at Kew

David V. Field

Richard Spruce's economic botany collections at Kew

David V. Field

Royal Botanic Gardens, Kew, Richmond, Surrey, UK

In 1849 Richard Spruce began 15 years of exploration in South America. On the 13th July 1849, the day after his arrival at Pará, he dispatched a letter to Sir William Hooker, the Director of Kew. Hooker and the eminent botanist George Bentham were prominent among a group of Spruce's acquaintances who had encouraged him to undertake the botanical exploration of the Amazon region, which was to become his life's work. Having no other source of income, this venture was to be financed by the sale of herbarium specimens, at the rate of £2 per hundred, which were to be sorted, authenticated, described when unknown and distributed to subscribers by Bentham. In return for acting as Spruce's agent Bentham received the first complete set of plants and these specimens became part of the Kew Herbarium. Hooker promoted Spruce's new career by publishing interesting extracts from correspondence received and emphasising the value and novelty of the collections. Thus a firm link was forged between Kew and Spruce which resulted in thousands of specimens, many of them new to science, being sent to the Royal Botanic Gardens at Kew.

Spruce travelled through dangerous and unexplored regions, often alone, but he met and befriended the local people, including Indian tribes, in these remote forests. He always took a lively interest in everything around him, noting the way of life and customs of the local people, their knowledge of the natural resources of the forest and how they made use of them. As well as voucher herbarium material of native economic plants, Spruce sent material for Kew's Museum of Economic Botany, opened only a few years before his departure. A list of specimens is given as Appendix A. Specimens transferred to the British Museum and the Pitt-Rivers Museum are listed in Appendices B and C respectively.

An early consignment of specimens destined for the Kew Museum is mentioned in Spruce's letter to Hooker dated April 1850. At this time his collecting was confined to the area of Santarém and he mentions that the only native manufacture of the town was baskets made from the Tucuma palm. The basket he sent is now in the British Museum, transferred in December 1960, together with other Spruce material. In his letter Spruce writes enthusiastically about commissioning a number of beautiful painted cuyas, bottle gourds, made only at Monte Alegre. In an earlier letter, 15 November 1849, Spruce gives an extremely detailed account of his observations on the remarkable giant waterlily, *Victoria amazonica*. Leaves and flowers preserved in a barrel of rum were noted as being received at the Kew Museum on 27 June 1850 but this delectable item is no longer present in the collections.

The main bulk of Museum specimens from Richard Spruce was sent to Kew between 1850 and the end of 1855, although smaller consignments continued to arrive throughout his stay in South America and even after his return. Samples of economic products included fruits, foods, fibres, gums and resins, oils, dyes, timber and drugs, both medicinal and narcotic. In addition, he sent to Kew numerous artefacts of clothing, weapons, ornaments and items both utilitarian and ceremonial. Some of the more spectacular items were those received at Kew on 12 March 1853. These are numbered by Spruce Nos. 122 to 143 and described by him as instruments and ornaments worn by the Uaupé Indians and principally by their chiefs, or Tuchánas during their festivals (called Dabocurís). Occasionally Spruce allows himself a brief whimsical comment, as in his description of a comb, No. 151, which is "used for combing out the long hair as well as for hunting the "Kínas" which always abound there".

The importance of the knowledge of local people, transmitted orally from generation to generation, encouraged him to record as much as he could, often in very difficult circumstances, in order to safeguard it for the future. Having been forced to return to England by increasing ill health, Spruce continued to document his uniquely acquired knowledge and a short extract from a letter to Hooker of 25 June 1864 shows us something of his indomitable spirit, enthusiasm and sense of humour in adversity: "I am waiting to see if I become a little stronger, in order to be able to walk about and see everything before I go over to Kew, and I am waiting also for a new masticatory apparatus from the dentist, for (like Bates and Wallace) I have lost most of my teeth in the forests of the tropics.

I have brought for your Museum, from the cotton-region of the north of Peru, specimens of several sorts of cotton, with a statement of their yield (weight of cotton and seed and cotton contained in an average pod of each; and weights of clean cotton, as left by the gin, of each sort). It is suggested to me to send a paper on the subject to the British Association. If I do so I cannot give up to you the specimens until after they have served to illustrate the paper".

The cultures which Richard Spruce encountered have diminished in the last hundred years or so, but in some more remote areas they still exist with little influence from the modern world. Recent ethnobotanical studies by Professor G.T. Prance, among others, have built on the foundations established by Spruce and added complementary items to the Kew collections. The tempo of change is quickening and the fragile ecosystems of the Amazon forests and the cultures of the indigenous people are threatened by increasing developmental changes. Much remains to be done to adequately record the botanical riches of these regions, to encourage their sustainable use and make clear the long-term benefits of so doing. Time is short to effect this rescue mission and the Royal Botanic Gardens Kew is ready to assist in this vital undertaking, as it was in the initiation of Richard Spruce's pioneering work.

Items displayed at the commemorative conference at York

The following notes are taken from contemporary listings in the Museum Entry Books which were extracted from Richard Spruce's letters and other sources. In most cases up to date Latin names are also given and further reference to these can be found in the main list (Appendix A).

Fig. 1. Apparatus for making and taking Niopo snuff (*Anadenanthera peregrina* (L.) Speg.), used by the Guahibo Indians. (Spruce 177).

1. Ambaúba or drum, of the trunk of *Cecropia peltata,* used by the Indians of São Gabriel in their Dabocurís or festas. It has been hollowed out by means of fire and the lower end closed with fresh leaves beaten hard down with a pestle. The performers in the dance beat them on the ground in unison with the movement of their feet. (When the leaves decay and fall out, the drum no longer gives its proper sound.) R. Spruce 152 Kew Cat. No. 43679.

2. Apparatus for making and taking Niopo snuff, procured from Guahibo Indians at the Cataracts of Maypures. (The Niopo of Venezuela is the same as the Paricá of Brazil, and is used on the Upper Orinoco, Guaviare, Vichada, Meta, Sipapo, etc. There is no doubt of its being prepared from *Acacia niopo* Humb., which is perhaps not different from *Piptadenia peregrina* Benth. [*Anadenanthera peregrina* (L.) Speg.].My specimens of the Paricá tree from Barra are referred to the latter species by Mr Bentham. I did not see the tree from which the Guahibos obtained their Niopo and which they told me was planted in their canncos near the headwaters of the river Tupáro; but the Paricá I have seen on the Amazon and all the way up the Rio Negro, planted near the villages, belongs to but one species, which, on passing the Venezuelan frontier takes the name of Niopo.)

    In preparing the snuff, the roasted seeds of the Niopo are placed in a shallow wooden platter, which is held on the knees by means of a broad handle grasped in the left hand, then crushed by a small pestle of the hard wood of the Pao d'Arco (*Tecoma sp.*).

The snuff is kept in a "mull" made of a tiger's bone, closed at one end with pitch and at the other stopped with a cork of marúna. It hangs from the neck, and has attached to it the tubiferous rhizomes of some Cyperaceae (*Hypoporum nutans* Nees?) which are slightly odoriferous. These or the tubers of some allied species are used throughout the Amazon, Rio Negro, Uaupés, etc. among the Indians of the forest. With a piece of Piripirióca (the name given to them in Lingua Geral) about the person, one is safe from bad wishes and evil eye.

The instrument for taking the snuff is made of bird's bones, and differs somewhat from that used by Catauixi Indians (see *J. Bot.* **5**: 246). Two tubes end upwards in little black balls (the endocarps of some species of *Astrocaryum*) which are applied to the nostrils while the single tube in which they unite at the lower end is dipped into the "mull", and thus the Niopo is snuffed up the nose.

I enclose a piece of Caápi, from which the Indian who was grinding the Niopo every now and then tore a strip with his teeth and chewed with evident satisfaction. It had been slightly toasted over the fire. "With a chew of Caápi and a pinch of Niopo", said he to me, in imperfect Spanish, "one feels so good, no hunger - no thirst - no tired!" A piece of Caápi is generally suspended along with the snuff box, but the snuffer or snuff-taker is stuck in the thick bushy hair of the head. R. Spruce 177, Kew Cat. No. 59120.

3. Arrows, such as the Tapuyas use at Santarém for killing fish. *Gynerium* sp. R. Spruce 38, Kew Cat. No. 38638.

4. Burnt bark of Caraipé. *Licania octandra* (Roemer & Schultes) Kuntze. R. Spruce 9, Kew Cat. No. 57006.

5. Bark. *Licania octandra* (Roemer & Schultes) Kuntze. Kew Cat. No. 57005.

6. Brooms of Piassába fibre. *Leopoldinia piassaba* Wallace. R. Spruce 102, Kew Cat. No. 35783.

7. *Cinchona pubescens* Vahl. Specimen of Red Bark. Ecuador, from the foot of Chimborazo, at 3000 feet altitude. Kew Cat. No. 52323.

8. Comb. This is worn stuck into the back hair of the head along with the tail (131). The teeth are of the stem of the Bacaba palm; they are inserted between the two masses of monkey's hair cord which are encased in slender strings of the culm of *Gynerium saccharoides*, interwoven with threads of Curaná. The free ends hang down the back and are ornamented at the extremity with parrot feathers. R. Spruce 132, Kew Cat. No. 34922.

9. Fish tongues used to grate Guaraná (*Paullinia cupana* Kunth); fish - *Arapaina gigas* (Piraracu). Kew Cat. No. 62374.

10. Fruits. *Paullinia cupana* Kunth. R. Spruce 71 Kew Cat. No. 62379.

11. Fruits of Guaraná. *Phytelephas macrocarpa* Ruíz & Pavón. Kew Cat. No. 36134.

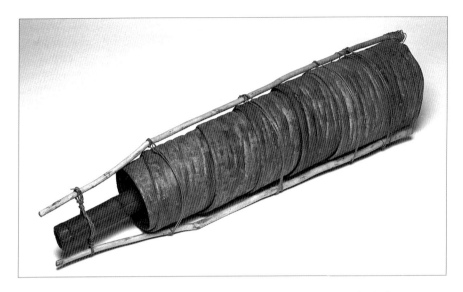

Fig 2. Jurupari, sacred trumpet made from *Iriartea exorrhiza* Mart. wrapped in bark of *Eperua grandiflora* Benth. & Hook.f. (Spruce 173).

12. Jurupari (or devil) used by the Indians on the Uaupés in their Dabocurís (festas). This is a musical instrument. It consists of a tube of the Paxiuba palm (*Iriartea exorrhiza*) wrapped with a long strip of the tough bark of the Jébarú (a Caesalpineous tree with handsome red monopetalous flowers, apparently the *Parivoa grandiflora* of Aublet), which descends in widening folds to some distance below the tube, thus forming a sort of trumpet, which is simply blown into at the upper end.

   I cannot find that the Juruparís are objects of actual adoration, but they certainly are of fear and respect. No woman is ever permitted to see them, and should such a circumstance occur, the woman is put to death, generally by poison, though the sight should have been accidental on her part. Youths are not permitted to handle or blow the Jurupari before the age of puberty, and must previously have undergone a series of fastings and scourgings.

   The Juruparís are kept hidden in the bed of some stream deep in the forest, in which no one dares to drink or bathe; and they are brought out only by night and blown outside the house where the feast is held, in order that no woman may obtain a sight of them. R. Spruce 173, Kew Cat. No. 59672.

13. Mandiocca grater made on the Rio Içanna, which enters the Rio Negro from the east, a little way above the mouth of Uaupés. Made of the soft tenacious wood of an Apocyneous tree (2265 to Benth.). The stones are chiefly a bluish fine-grained granite from the Içanna, broken into fragments of convenient size. Design scratched with point of a large nail; then with the same a hole is pricked for the insertion of each stone, and a blow of the nail head secures it in its place. The grater thus formed is anointed with the milk of the Cumá (*Couma dulcis* Benth. and other Apocynaceous trees probably of the same genus) which is a powerful adhesive, not affected by juice of Mandiocca or

Fig. 3. Mandiocca grater with granite teeth attached with sap of *Couma dulcis* Muell. Arg. (Spruce 189); tapioca flour prepared from *Humirianthera rupestris* Ducke (Spruce 80).

other moisture. I have seen graters which were decayed and almost worn through at the back, while not a tooth had fallen out. Içanna graters are in great request throughout the Rio Negro and Amazon, and even on the Orinoco. R. Spruce 189, Kew Cat. No. 38689.

14. Mandiocca strainer of Marantaceous petiole. *Ischnosiphon* sp. R. Spruce 30 1849, Kew Cat. No. 38688.

15. Ollitas. (Small jars). *Phytelephas macrocarpa* Ruíz & Pavón. Kew Cat. No. 36045.

16. Paddle made at Santarém from Itauba amarella (*Acrodiclidium ita-uba*). R. Spruce 37, Kew Cat. No. 45487.

Fig. 4. Pánela made from burnt bark and clay; bark of Pottery tree, *Licania octandra* (Roemer & Schultes) Kuntze. (Spruce 19/1849).

17. Pénalas, used for heating milk, boiling eggs, and other purposes. Made from equal parts of clay and the bark of Caraipé or Pottery tree, *Licania octandra* (Roemer & Schultes) Kuntze. R. Spruce 19/1849, Kew Cat. No. 37800.

18. Poisoned Arrows or Curabí of the Uaupés (Maupe Indians). *Gynerium saggitatum* (Aublet) Pal. R. Spruce 143, Kew Cat. No. 38648.

19. Pottery ware, probably made on the River Uaupés from the bark of one or more species of *Licania* or *Moquilea* mixed with clay. Kew Cat. No. 55451.

20. Quiver (called Marupá) belonging to Gravatano or blowing cane made by Catauixi Indians on the Rio dos Purús and containing poisoned arrows. It is made of the Sipo Oambe-cîma (the wood not the bark) and thickly smeared with the resin of the Jutahí (*Hymenaea* sp.). The arrows are made of the nerves of the sheathing base of the petioles of the Pataná, which, remaining when the parenchyma decays, forms a sort of beard on the trunk. They are called by the Catauixi Indians Araráicohi and the poison with which they are smeared Arimilihá (Uirarí). It is customary to make them up into bundles such as I send you. When an Indian goes a-hunting he takes out as many arrows as he is likely to want, anoints the points with Uirarí and wraps the lower end with Simaüma-cotton to the thickness of the bore of the Gravatano. R. Spruce 77, Kew Cat. No. 35161.

21. Resin. *Hymenaea courbaril* L. Kew Cat. No. 59746.

Fig. 5. Anklet of *Cayaponia kathematophora* R. Schultes seeds worn by Barré Indians (Spruce 158); comb made from *Oenocarpus bacaba* Mart. and *Gynerium sagittatum* Mart. (Spruce 132).

22. Shells of some fruit (*Cayaponia kathematophora*) strung together, and tied round the right ankle in the Dabocurís (dances) of the Barré Indians, producing a loud rattling noise with every movement of the wearer. They come from the Rio Içanna. R. Spruce 158, Kew Cat. No. 35009.

23. Shield of sipo called Timbo-titica (Menisperm. ?). It is partially smeared with pitch of Anani (*Symphonia globulifera*). R. Spruce 140, Kew Cat. No. 67762.

24. Shirt of Tururí. Called Tácaé by the Cubén Indians on the Rio Uaupés, who use them in their funeral feasts, when they drink the ashes of the bones of their deceased relatives.

    (There are two sorts of Tururí; the common red, which is the bark of a large Artocarpaceous tree, allied to *Antiaris*, frequent on the Rio Negro and Casiquiari, and of which I procured specimens at São Gabriel; it is used for bags, for caulking canoes, and on the Guainia and Casiquiari (where it is called Maríma) a rude kind of shirt is made of it. The Tururí-morotinga, or white Tururí, of which the bodies of the Tácaé are made, is the bark of a real fig (*Ficus* sp.), a low terrestrial species, which I could not distinguish by the leaves from a species I had gathered near São Gabriel).

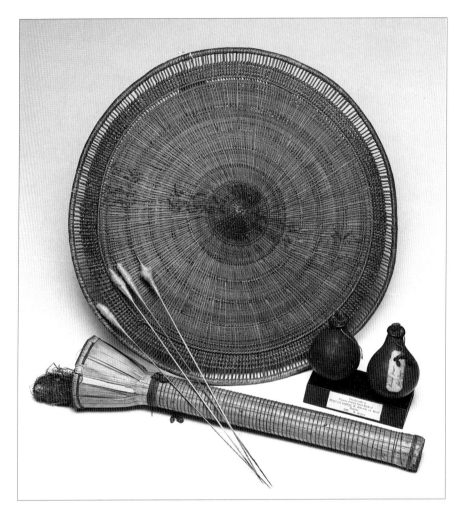

Fig. 6. Shield smeared with pitch from *Symphonia globulifera* L. f. (Spruce 140); quiver and curare poisoned arrows made by Catauixi Indians (Spruce 77); taparitós (gourds) containing curare made by Indians on Rio Pacimoni (Spruce 184).

The arms of the Tácaé are of red Tururí; the fringes of Sapucaya Castanha (a name applied to all large fruited Lecythideae). The colours used in painting them are carajurú or arnatto for red and soot for black. R. Spruce 174, Kew Cat. No. 43397.

25. Small bucket used on board canoes on the Amazon and Rio Negro. It is merely a hollowed cuya (calabash - *Crescentia cujete*) with a handle of piassaba (*Attalea funifera*) attached to two crossed bands of netted curana cloth. R. Spruce 149, Kew Cat. No. 46534.

26. Taparitós (small gourds) of Curári (the Uirarí or"bird poison"of Brazil). Made by the Indians of the river Pacimoni from the bark of two downy leaved-

species of *Strychnos*. I fear the poison will be quite dried up by the time it reaches you. The Indians keep it in a cool moist place, and if it becomes stiff, set the Taparatós for some time in the moist ground, or boil the curári over again. R. Spruce 184, Kew Cat. No. 49120.

27. Tapioca prepared from Bauná Root. *Humirianthera rupestris* Ducke. R. Spruce 80, Kew Cat. No. 62880.

APPENDIX A

List of specimens collected by Richard Spruce in the collections of Economic Botany, Kew, 1993

This list is in alphabetical order of genus and species names, except for 7 unidentified specimens at the end. Spruce's numbering of specimens sent for the Kew Museum was not consistent. His first 3 boxes contained specimens numbered 1 to 31; we have no record of 32-34 but No. 35, received on 27 June 1850, was the barrel containing *Victoria amazonica* in rum. The consignment received on 22 July 1850 is numbered 1 to 29, and this sequence continues, with some miscellaneous interpolations, to No. 196 received in October 1856. No. 81 consists of 22 different woods, most of which are identified by corresponding herbarium numbers, while No. 82 consists of 8 specimens of different fruits in spirit. Some of the miscellaneous additions have identifying herbarium specimen numbers but most of those subsequent to No. 196, consisting principally of woods, flowers, fruits and seeds, do not have any Spruce numbers.

Kew Catalogue Number		Spruce Number
19705 & 19706	Wood of *Abuta concolor* Poepp. & Engl.	1251
470 & 471	Wood of *Abuta rufescens* Aubl.	148=2192
58582	Pods of *Acacia farnesiana* Willd. Peru, Tarapoto	4548
2994	Sack of fibre of *Agave* sp.	179
59926	Fruits of *Aldina latifolia* Spruce & Benth. Barra Amazon, São Gabriel	2007
10949	Wood of *Amaious saccifera* Mart.	-
52233	Fruits of *Amaious velutina* Spruce	-
49532	Fruits of *Ambelania acida* Aubl. Brazil, Rio Negro (Lectotype)	-
61857	Gum of *Anacardium occidentale* L. Brazil, Santarém	52
58929	Pods of *Anadenanthera peregrina* (L.) Speg. Brazil, Rio Negro	119
59120	Apparatus for making and taking Niopo snuff *Anadenanthera peregrina* (L.) Speg. Colombia, Maipures, Rio Orinoco	177
45652	Twisted stem of *Aristolochia* sp. Brazil	-

45645	Fruit of *Aristolochia* sp. Peru	4206
45630	Fruits of *Aristolochia* sp. Peru, Tarapoto	-
46475	Chica, *Arrabidea chica* (Bonpl.) Verl., prepared by Indians for adornment. Brazil, Rio Negro	65
466466	Red dye of *Arrabidea chica* (Bonpl.) Verl., in bag of bark. Brazil, Rio Negro	65
49540	Fruits of *Aspidosperma* cf. *desmanthum* Benth.	3393
49553	Fruits of *Aspidosperma pachypterum* Muell. Arg.	3712
26221	Wood of *Astrocaryum gynacanthum* Mart. Brazil, Santarém	31
40022	Spadix of *Astrocaryum vulgare* Mart. Brazil, Pará	4/1849
35010	Rope of *Astrocaryum vulgare* Mart. Brazil, Barra de Rio Negro	66
35009	Rings made of seeds of *Astrocaryum vulgare* Mart. Brazil, São Gabriel	107
34980 & 35008	Fruits of *Astrocaryum vulgare* Mart. Brazil, São Gabriel	154
40588	Quiver of *Attalea* sp. Brazil, Rio Uaupés	143
38730	Spathe of *Attalea* sp.	-
35014	Fruits of *Attalea spectabilis* Mart. Brazil, Santarém	32
35099	Fruits of *Bactris acanthocarpa* Mart. Brazil, Tanari	10/1849
35072	Fruits of *Bactris bidentula* Spruce. Brazil, Rio Negro	91
35071	Fruits of *Bactris bifida* Mart. Brazil, Rio Negro	109
35092	Fruits of *Bactris concinna* Mart. Brazil, Rio Negro	110
38867	Fruit and spadix of *Bactris concinna* Mart. Brazil, Rio Negro	92
35088	Female spadix of *Bactris maraja* Mart. Brazil, Pará	5/1849
67428	Stems of *Banisteria* sp. Brazil, Rio Uaupés	166=2712
59282	Stem of *Bauhinia* sp. Brazil, Rio Uaupés	99
55027	Bark of *Bertholletia excelsa* Bonpl. Brazil, Pará	25/1849
65263	Fruits, seed and seed fibre of *Bombax munguba* Mart. Brazil, Santarém	50
29727	Fibre and rope of *Bromelia* sp. Brazil, Santarém	42
42857	Bark of *Brosimum* sp. Brazil, Rio Uaupés	146
64471	Bark of *Byrsonima spicata* Rich. Brazil, Rio Para, Caripi	20/1849
64470	Bark of *Byrsonima spicata* Rich. Brazil, Rio Uaupés, Rio Para, Carya	
59409	Pods of *Caesalpinia spinosa* (Mol.) Kuntze	5315
66493	Fruits of *Caryocar glabrum* (Aubl.) Pers. Brazil, Rio Negro	253
52265	Bark of *Cascarilla* sp. Ecuador, Quitonian Andes	198
11009	Wood of *Cascarilla magnifolia* Wedd. Peru, Chimborazo	-
59525	Pods of *Cassia leiandra* Benth.	-

54476	Anklet of seeds of *Cayaponia kathematophora* R. Schultes. Brazil, Rio Negro	158
43679	Ambaúba or drum of *Cecropia peltata* L. Brazil, São Gabriel	152
16239 & 16242	Wood of *Cecropia scabra* Mart.	1322
65316 & 65317	Fruits of *Ceiba samauma* K. Schum. Brazil, Rio Negro	117
65318	Silk-cotton or kapok of *Ceiba samauma* K. Schum.	-
56978	Fruits of *Chrysobalanus icaco* L. Brazil, Rio Aripecurú	6
50793	Fruits of *Chrysophyllum* sp. Brazil, Rio Negro	82=1393
50803	Bark of *Chrysophyllum* sp.	81c=1393
52508	Bark of *Cinchona pubescens* Vahl. Ecuador, Quitonian Andes	-
52322	Bark of *Cinchona pubescens* Vahl. Ecuador, Forests of Chimborazo	4/1861
52323	Bark of *Cinchona pubescens* Vahl. Ecuador, foot of Chimborazo	-
52496	Fruits of *Cinchona pubescens* Vahl. Peru	-
11024	Wood of *Cinchona pubescens* Vahl. Ecuador, Forests of Chimborazo	-
53605	Fruit of *Condaminea corymbosa* DC.	4579
56988	Fruits of *Couepia racemosa* Hook.f. Brazil, Barra de Rio Negro	2322
49632	Concrete milk of Cow Tree, *Couma* sp.	-
49626	Mandiocca grater. *Couma utilis* Muell. Arg. Brazil, Rio Içanna, tributary of Rio Negro	189
3767	Wood of *Crepidospermum rhoifolium* Benth.	1344
46534	Small bucket of calabash, *Crescentia cujete* L. Brazil	149
48619	Stem of *Cuscuta racemosa* Mart. Brazil, Santarém	16
46533	Fruits of *Cybistax antisyphilitica* (Mart.) DC. Peru	4269
41040	Fruits of *Cymbopetalum brasiliense* Benth.	4097
50606	Fruits of *Diospyros peruviana* Hiern. Brazil	4411
60295	Fruits of *Diplotropis* sp. Brazil, Pará	-
60300	Fruit of *Dipteryx odorata* (Aubl.) Willd. Brazil, Pará	29/1849
60313	Fruit of *Dipteryx* sp. Brazil, Barra de Rio Negro	64
6347	Wood of *Dipteryx tetraphylla* Benth. Brazil	-
49656	Follicles of *Echites stellaris* Lindl.	4900
35663	Male spadix of Caiane, *Elaeis oleifera* (Kunth) Cortes. Brazil, Rio Negro	74
53636	Female spadix with fruits of *Elaeis oleifera* (Kunth) Cortes. Brazil, Rio Negro	111
35633	Fruits of *Elaeis oleifera* (Kunth) Cortes.	-
62879	Fruits of *Emmotum orbiculatum* Miers.	3541

59672	Juruparí (or devil), sacred musical instrument made from *Iriartea exorrhiza* Mart. and *Eperua grandiflora* Benth. & Hook.f. Brazil, Rio Uaupés	173
35655	Fruits of *Euterpe catinga* Wallace.	-
35673	Spadices of *Euterpe edulis* Mart. Brazil, Pará	3/1849
26283	Wood of *Euterpe* sp.	113
35629	Fruits of *Euterpe* sp.	-
54601	Seeds of *Fevillea cordifolia* L.	-
54607	Fruit of *Fevillea* sp.	4031
54606	Fruit and seeds of *Fevillea* sp.	-
43397	Shirts worn by Indians at funeral feast. *Ficus* sp. Brazil, Rio Uaupés	174
60379	Fruits of *Geoffroea striata* (Willd.) Morong. Ecuador, sea coast near Guayaquil	-
72564 & 35683	Female spadix with fruits of *Geonoma baculifera* (Poit.) Kunth. Brazil, Pará	-
27112	Fruits of *Gnetum paniculatum* Spruce	-
27115	Fruits of *Gnetum* sp.	-
65450	Kidney cotton and seeds. *Gossypium peruvianum* L. Peru, Rio de la Chira, Monte Abierto	-
65535	Ica cotton and seeds, *Gossypium* sp. Peru, Rio de la Chira, Monte Abierto	-
65535	Cotton capsule contents of *Gossypium* sp. Peru, Rio de la Chira, Monte Abierto	-
65469	Imbabura cotton, *Gossypium* sp. Peru, Rio de la Chira, Monte Abierto	-
65515	Contents of 3 celled cotton capsule. *Gossypium* sp. Peru, Rio de la Chira, Monte Abierto	-
65492	Cotton, *Gossypium* sp. Peru, Rio de la Chira, Amatope	-
65600	Guayacu of stout cotton cloth, *Gossypium* sp. Venezuela, Orinoco	175
38648	Poisoned arrows of *Gynerium saggitatum* (Aubl.) Beauv. Brazil, Rio Uaupés	143
31969	Plant of *Gynerium saggitatum* (Aubl.) Beauv. Venezuela, Rio Casiquare, Pueblos de Monagas	-
38727	Part of stem of *Gynerium* sp. Brazil, Amazon, Ilha-de-Mari-Mari-tuba	34
38638	Arrows of *Gynerium* sp. Brazil, Santarém	38
1240 to 1242	Wood of *Haploclathra paniculata* Benth. Brazil, Santarém	46
15882	Wood of *Hevea guianensis* Aubl. Brazil, Pará, Belém	-
44064	Fruit and seeds of *Hevea spruceana* Muell. Arg.	3289

44172	Drumstick of *Hevea* sp. Brazil, Rio Uaupés	171
15885	Wood of *Hevea sp.* Brazil	-
49721	Fruit and seeds of *Himatanthus phagedaenica* (Mart.) Woods. Brazil, Santarém	14
62781	Fruits of *Hippocratea megacarpa* Peyr. Brazil, Rio Uaupés, Pará. Washed on shore at Ciripi	2796
64479	Bark of *Humiria floribunda* Mart. Brazil, Pará	13/1849
62882 & 62883	Root of *Humirianthera rupestris* Ducke. Brazil, Barra de Rio Negro	80
62880 & 62881	Farina and tapioca prepared from Bauna root, *Humirianthera rupestris* Ducke. Brazil, Rio Negro	80
59747	Resin of *Hymenaea courbaril* L. Brazil, Santarém	54
4681	Wood of *Ilex parviflora* Benth. Brazil	1141
62858	Leaves of *Ilex* sp. Ecuador	-
35742	Fruiting spadix of *Iriartea exorrhiza* Mart. Brazil, Pará	9/1849
35741	Fruiting spadix of *Iriartea exorrhiza* Mart. Brazil, Rio Negro	-
35719	Seeds of *Iriartea exorrhiza* Mart. Brazil, Rio Negro	112
26296	Wood of *Iriartea pruriens* Spruce. Brazil, Rio Negro	-
35785	Spadix of fruits of *Iriartea ventricosa* Mart. Brazil, São Gabriel	153
38688	Mandiocca strainer made from *Ischnosiphon* sp. petiole. Brazil	30/1849
51194	Fruits and seeds of *Jacquinia sprucei* Mez. Ecuador, sea coast near Guayaquil	-
35729	Poisoned arrows made from spines of *Jessenia bataua* (Mart.) Burret	-
35161 & 35776	Quiver and poisoned arrows of *Jessenia bataua* (Mart.) Burret. Brazil, Rio dos Purús	77
35786	Part of spadix of Patana palm, *Jessenia* sp. Brazil, Barra de Rio Negro	87
54813	Oil bottle of *Lagenaria siceraria* (Molina) Standley, cased in basket work. Venezuela, Rio Cunucunuma	188
71802	Maraca of *Lagenaria siceraria* (Molina) Standley. Brazil, Rio Uaupés	-
54686	Seeds of gourd, cf. *Lagenaria siceraria* (Molina) Standley, used as beads. Brazil, Rio Amazon	-
10281	Wood of *Lecythis* sp. Brazil, Pará	-
55077 & 55084	Bark of *Lecythis* sp. Brazil, Pará	27/1849
55058	Bark of *Lecythis* sp. Brazil	47
26298	Wood of *Leopoldinia pulchra* Mart.	-
35766	Fruits of *Leopoldinia piassaba* Wallace. Venezuela, Casiquiare	-

26300	Wood of *Leopoldinia piassaba* Wallace	-
35779	Piassaba fibre, *Leopoldinia piassaba* Wallace	191
35783	Brushes or brooms of Piassaba fibre, *Leopoldinia piassaba* Wallace. Brazil, Rio Negro	102
35838	Petioles of *Leopoldinia piassaba* Wallace	191
35761 & 35764	Fruits of *Leopoldinia pulchra* Mart. Brazil, Rio Negro, Triquim & Santarém	84
57006	Ashes of bark of *Licania octandra* (Roemer & Schultes) Kuntze. Brazil, Pará	9
57005	Bark of *Licania octandra* (Roemer & Schultes) Kuntze Brazil, Pará	19/1849
37793	Leaves of *Licania octandra* (Roemer & Schultes) Kuntze Brazil, Pará	19/1849
37794, 37796, 37799 & 37800	Bowls/Panelas made from bark or *Licania octandra* (Roemer & Schultes) Kuntze and clay. Brazil, Pará	19/1849
37798	Model of Fogareiro or chafing dish made of bark/ashes of *Licania octandra* (Roemer & Schultes) Kuntze and clay. Brazil, Pará	-
55451	Pottery ware made from bark/ashes of *Licania* sp. Brazil, Rio Uaupés	-
57009	Fruits of *Licania* sp. Brazil	-
44765	Stem with *Loranthus* sp. showing germination	-
57015	Fruits of *Lovepia glandulosa* Miq. Brazil	2262
54707	Fruit of *Luffa aegyptiaca* Miller	10
44291	Tobacco pipe of *Mabea fistulifera* Mart. Brazil, Pará	22/1849
44373	Fruits of *Manihot peruviana* Muell. Arg. Peru	4287
26314	Wood of *Mauritia aculeata* Kunth. Brazil, Santarém	49
35856	Fruits of *Mauritia armata* Mart. Brazil, Rio Negro	-
35911	Fruits of *Mauritia flexuosa* L. f. Brazil, Pará	1/1849
6318	Wood of *Mauritia flexuosa* L. f. Brazil, Pará	1/1849
35859	Fruits of *Mauritia flexuosa* L. f. Brazil, Pará	1/1849
38775	Spadix of *Mauritia martiana* Spruce. Brazil, Santarém	49
35912	Fruits of *Mauritia martiana* Spruce. Brazil, Santarém	-
26317	Wood of *Mauritia vinifera* Mart. Brazil, Pará	2/1849
35857	Fruits and seeds of *Mauritia* sp. Venezuela, San Carlos	-
45048	Bark of *Mezilaurus ita-uba* Mez. Brazil, Santarém	18
45487	Paddle of *Mezilaurus ita-uba* Mez. Brazil, Santarém	37
50915	Stem of *Mimusops elata* Allem. Brazil, Pará	21/1849
50890 & 50892	Concentrated milk of *Mimusops elata* Allem. Brazil, Pará	-
12063	Wood of *Mimusops longifolia* A. DC. Brazil, Minas Geraes, Pará	-

1326	Wood of *Moronobea globulifera* Schlecht. Brazil, Caripi	-
17184	Wood of *Myrcia barrensis* Benth. Brazil, Barra de Rio Negro	part 81=1341
42560	Wax from berries of *Myrica* sp. Ecuador, Forest of Canelos	-
6549	Wood of *Myroxylon balsamum* (L.) Harms. Ecuador, Chongon, near Guayaquil	-
58151	Bark of *Myroxylon robiniaefolium* Klotsch. Ecuador, Chongon, near Guayaquil	-
14960	Wood of *Ocotea obscura* Meisn.	-
34922	Comb made from *Oenocarpus bacaba* Mart. and *Gynerium sagittatum* (Aublet) Pal.	132
35969	Fruits of *Oenocarpus bacaba* Mart. Brazil, Rio Negro	87
35982	Spadix of *Oenocarpus minor* Mart. Brazil, Barra de Rio Negro	89
35968	Stem and spadix of Bacaba palm, *Oenocarpus minor* Mart. Brazil, Barra de Rio Negro	89
35981	Portion of Curoa palm, *Orbignya spectabilis* (Mart.) Burr. Brazil, Santarém	32
60837	Pods and seeds of *Ormosia coccinea* (Aubl.) Jackson. Venezuela, San Carlos	-
355	Wood of *Oxandra multiflora* Mart.	part 81=1301
60861	Pods and seeds of *Pachyrrhizus tuberosus* (Lam.) Spreng. Peru, Tarapoto	4936
59205	Pods of *Parkia oppositifolia* Benth.	-
59098	Pods of *Parkia pendula* Benth. Brazil, Santarém	-
62379	Fruits of *Paullinia cupana* Kunth. Brazil, Rio Negro	71 & 118
62373	Guarana, *Paullinia cupana* Kunth. Brazil, Rio Manhe	39
30343	Fruiting body of *Phellinus* sp. Brazil, Panure (São Jeronymo)	-
38674	Tough pendulous roots of *Philodendron goeldii* G.M. Barroso. Brazil, Rio Negro	120
36134	Fruits of *Phytelephas macrocarpa* Ruíz & Pavón. Brazil, Rio Purus	-
36045	"Ollitas", model jars made of vegetable ivory, *Phytelephas macrocarpa* Ruíz & Pavón	
49947	Wooden ladle of *Plumeria mulongo* Benth.	-
47907	Fruits of *Poraqueiba sericea* Tul. Brazil, Barra de Rio Negro	part 82
51117	Gum of *Pouteria sapota* (Jacq.) H.E. Moore	-
49955	Roots of *Prestonia* sp. Brazil, Rio Uaupés	part 166
46416	Fruits of *Proboscidea peruviana* Van Es. Peru, Desert of Sechura	-
63396	Block of resin of *Protium* sp. Venezuela/Colombia, Rio Orinoco	185
79491	Wood of *Protium spruceanum* Engl. Brazil, Rio Negro	-

53805	Fruits of *Randia* sp.	4905
62411 & 62451	Fruits and seeds of *Sapindus saponaria* L. Brazil, Santarém	51
53856	Fruits of *Schradera spicata* Hook. f.	3322
62441	Roots of *Serjania curassavica* Radlk. Brazil	-
51293	Caoutchouc of *Siphocampylus jamesonianus* A. DC. Ecuador, Baños	-
36987	Stem of *Smilax* sp. Brazil, Barra de Rio Negro	98
36986	Stem of *Smilax* sp. Brazil, Rio Negro	160
36335	Stem of *Socratea* sp.	-
48530	Fruits of *Solanum* sp. Brazil, Rio Negro	82=1347
13519	Wood of *Solanum* sp.	part 81=1347
48519	Fruits of *Solanum* sp. Brazil, Rio Negro	233
62147	Fruits of *Spondias mombin* L. Brazil, Rio Negro	-
62148	Gum of *Spondias mombin* L. Brazil, Rio Negro	-
53898	Galls of *Stachyarrhena spicata* Hook.f.	-
49120	Two gourds containing curare, *Strychnos toxifera* Benth. Venezuela, Rio Yatua (Rio Pacimoni)	184
66685	Pitch/resin of *Symphonia globulifera* L. f. Venezuela/Colombia, Rio Orinoco	186
67762	Shield smeared with pitch of *Symphonia globulifera* L. f. Brazil, Rio Uaupés	140
66675	Gum of *Symphonia globulifera* L. f. in palm fruit. Brazil, Pará	14/1849
46604	Gigar support of *Tecoma* sp. Brazil, Rio Uaupés	142
56768	Fruits of *Terminalia punctata* Eichl. Peru	4945
65168	Fruits of *Theobroma martiana* D. Dietr. Brazil, Barra de Rio Negro	97
50059	Seeds of *Thevetia peruviana* (Pers.) K. Schum. Brazil, Marabitanas	2914
3856	Wood of *Trattinickia rhombifolia* Willd.	-
63104	Bark of *Trichilia* sp. Ecuador, Chimborazo	-
937 & 939	Wood of *Vochysia vismiaefolia* Warm.	-
36373 & 36374	Spadices of *Wettinia maynensis* Spruce. Peru	-
33442	Ears of Indian Corn, *Zea mays* L. with corn removed. Peru	-
38689 & 49626	Mandiocca (Cassava) graters. Brazil, Rio Içanna	-
45694	Salt made from various Podostemaceae species. Brazil, Rio Uaupés	164
57057	Fruits of Chrysobalanaceae. Brazil, Rio Tapajós	53
62374	Fish tongues used to grate guarana. Brazil, Amazon	-
62466	Wooden spoons and wood. Ecuador, Quitonian Andes	-

| 63403 | Resin. Brazil, Pará | – |
| 68833 | Fruits of Chrysobalanaceae. Brazil, São Gabriel | 161=2197 |

APPENDIX B

Spruce specimens and notes sent to the British Museum 21 December 1960

Spruce No. 24.
Tapuya work basket, made by the Indians at Santarém of the young leaves of the Tucumá Palm. 1850.

Spruce No. 62.
A piece of wood from which fire has been drawn by the Indian method. 1851.

Spruce No. 76.
Gravatano or Blowing cane made by Catauixi Indians on the Rio dos Purús. It is a portion of the trunk of the paxiuba-i (a small species of *Iriartea*, very near *I. setigera* Mart.) 9ft 3in long, wrapped round with the bark of a sipo called Oambé-címa. At about 2ft from the lower extremity is the sight, one or two teeth of the Cutía stuck on with resin of the Jutahí. 1851.

Spruce No. 101.
Hat made of the leaf of the Tucumá Palm, *Astrocaryum tucuma* Mart. 1851.

Spruce No. 104.
Maqueira (hammock) made by the Indians on the Rio Napo, of the young fronds of the Tucumá palm. February 1852.

Spruce No. 135.
Arm ornament of parrot feathers fastened to monkey's hair strings and meeting over a hollowed fruit of Tucúm (*Astrocaryum vulgare*), into the cavity of which a small pebble has been inserted. This is worn over the elbow. 12 March 1853.

Spruce No. 138.
Beads, worn shot-bag wise, over the left shoulder and under the right arm. 12 March 1853.

Spruce No. 139.
Box in which the above articles are contained. It is made of the pinnae of the frond of Uauassú (*Attalea* sp.) crossed by Tucúm string. The frame of the top and bottom seems to be paxiuba. 12 March 1853.

Spruce No. 156.
Maqueira (hammock) of Murití. The cord from which this is woven is made from the cuticle of the fronds of Murití palm, which is stripped off in the same manner as that of the Tucúm. Hammocks made of Murití are softer but less durable than those of Tucúm. 12 March 1853.

Spruce No. 157.
Petiole of Caraná assú (*Copernicia* sp.) with the skin stripped off, in which state it is used on the Rio Negro for corks, birdcages &c. 12 March 1853.

Spruce No. 167.

Ornamental hammock (called Maqueira in Brazil, Chinchorro in Venezuela) made at Tomo on the Guainia (Upper Rio Negro). The body of the hammock is made from the fronds of *Astrocaryum vulgare* (called Tucúm in Brazil, Cumári in Venezuela). The borders are an open network made from the fronds of *Mauritia vinifera* (called Murití in Brazil, Moríche in Venezuela); the white feathers with which they are ornamented are those of the Royal Heron, the black of the Curassow, and the rest are parrot, macaw, humming birds etc. The cords are Tucúm. 17 January 1854.

Spruce No. 170.

Indian "bellows", made of strips of the leafstalk of Tucúm. From the Rio Uaupés. 17 January 1854.

Spruce No. 176.

Two baskets of strips of rind of various species of *Maranta*, made by Maquiritare Indians on the Rio Cunucunuma, and used by them for holding their tinder box, fish hooks, arrow heads etc. 15 June 1855.

APPENDIX C

Spruce specimens and notes sent to the Pitt Rivers Museum 27 January 1961

Spruce No. 43.

Twine called "Maqueira" made of the fibre of the leaf of the Tucumá palm (*Astrocaryum vulgare* Mart.). Of this are made the hammocks which some travellers have called "Grass hammocks". 1850.

Spruce No. 58.

"Birros", used on the Amazon in making cushion lace. Seven of them are of the fruit of the Tucumá palm, the other five of the fruits of the Mucajá palm, four of the latter being polished by rubbing them with a stone. 7 March 1851.

Spruce 103.

Mat made at Myobamba in the Peruvian Andes of the young fronds of the Murití palm (*Mauritia* sp.). February 1852.

Spruce 131

Tail of monkey's hair cord. Worn hanging down the back, the loop being fixed over the deer's bone.

Spruce 137.

Two pairs of garters woven of Curana thread and painted with taná (yellow earth) and Carajurú. Children wear similar garters almost from infancy, and the leg just below the knee is so tightly compressed by them that a deep and permanent impression is produced. 12 March 1853.

22

Spruce in Manchester

Sean R. Edwards

Spruce in Manchester

Sean R. Edwards

*The Manchester Museum, Manchester University, Oxford Road,
Manchester M13 9PL, UK*

with an Appendix on Manchester City Library holdings, by Brian W. Fox

(Updated from the British Bryological Society lecture read at Ripon on 18 September 1993, and from the poster shown at Ripon and at the Linnean Society meeting at Castle Howard on 21 September)

Introduction

The large holdings of Richard Spruce material at Manchester Museum constitute about 16,500 items, and seem to have been one of the best kept secrets about this remarkable Yorkshireman who died one hundred years ago. Not only do we have Spruce's own large personal herbarium, with sets of his *Hepaticae spruceanae: Amazonicae et Andinae*, his *Musci Amazonici et Andini* (incomplete), and substantial lichen collections, but we also have a collection of his letters and maps and other documents which may interest Spruceologists more than bryologists.

Although Richard Spruce (1817–1893) was a Yorkshireman, he had many and continuing connections with the active network of bryologists on the western Lancashire side of the Pennines that centred on Manchester. The two great bryologists William Wilson (1799–1871) of Warrington (then in Lancashire), and John Nowell (1802–1869) of Todmorden (only 15 miles from Manchester, though just in Yorkshire), were nearly twenty years his senior. Spruce also corresponded with the younger Benjamin Carrington (1827–1893) who was President of the Manchester Cryptogamic Society, and whose own herbarium was bequeathed to Manchester University, as were those of many others including Nowell and John Whitehead (1833–1896) of Oldham (Lancashire).

In 1867, the very substantial collections of the Manchester Natural History Society (founded in 1821) were taken over by the Governors of Manchester University, then known as Owens College, thus maintaining the continuity and increasing the status of the collections and attracting further contributions. Therefore Manchester became the obvious repository for major cryptogamic collections from the North of England. George Stabler (1839–1910), once a pupil at Ganthorpe School where Spruce's father taught,[1] also sent his valuable herbarium over the hills to Manchester.

But it was Mancunian W.H. Pearson (1849–1923) who was ultimately responsible for the transfer of Spruce's massive personal herbarium to Manchester University in 1919, nearly 26 years after Spruce's death. Details of the Spruce acquisitions are given in this paper under the heading of **The Spruce Liverworts**. The University museum, known as The Manchester Museum, also has Pearson's own British Hepaticae.[2] The number of cryptogams, and bryophytes in particular, accumulated by the Museum at the end of that great period was not much less than the present figures of 270,000 and 135,000 respectively.

The Spruce letters and newspapers

The Spruce letters are in his own hand, mostly to Matthew B. Slater of Malton, Yorkshire; they start simply *"Dear Sir"*. The bulk is dated from 13 March 1880, to 20 September 1893, just two months before his death, with a few gaps such as most of 1881, and much of 1886. In these letters, Spruce asks Slater for items such as ink, bread, or spectacles, grumbles to him about other bryologists, makes penetrating asides on Victorian politics (he was Liberal), and recounts events from the mundane to the bryologically important. Slater was Spruce's old friend, factotum, confidant, and eventually his botanical executor; he lived some four miles from Coneysthorpe. The letters in effect form a diary of Spruce's last thirteen years. They make fascinating reading, both from a social and historical standpoint, and also for any bryologist interested in perceptive and detailed observations by one of the world's greatest hepaticologists. The letters have now been typed onto disc in their entirety (60,870 words), and have also been indexed on Spruce's (expanded) scientific names; indexing of other aspects is continuing. Selected extracts of the letters are given in Appendix A.

These letters to Slater are concurrent with those written to George Stabler, of which Wallace[3] says:

> *"The only records of the last fifteen years of Spruce's life which are available are in the continuous series of letters to his life-long friend Mr G. Stabler, who, both as schoolmaster, invalid, and botanist, was in complete sympathy with him".*

It is not clear why the Slater letters now at Manchester were *"unavailable"* to Wallace. Despite Wallace's frustrated letters to Slater (as evidenced by a letter from Wallace to Slater, dated 6 September 1906; see Appendix C: collection 3), Slater does seem to have made great efforts to assist Wallace, providing papers for him after Spruce's death; Slater was even *"asked to provide a list of friends with whom Spruce might have corresponded, and much seems to have been recovered"*.[4] Yet the only letter from Spruce to Slater mentioned by Wallace is that dated October 1851, from Barra do Rio Negro.[5] Why the Manchester Slater letters seem to have been seen only by Spruce, Slater, and a few personal visitors to the Museum, remains a mystery.

Sadly, however, we do not seem to have the famed *"missing journal"*, that so eluded Wallace.[6] At least this seems to add weight to Slater's protestations that he had never had it.

There are also some hand-written species descriptions, and other items with the letters, totalling over 300 items. In addition we have letters by Spruce to other people, for example his letters to Benjamin Carrington in the Carrington file. Of peculiar interest is a short poem to Miss W., dated April 19, 1869, about five years after his return from South America (Miss W. was possibly the Miss

Waller mentioned in Spruce's letter to Slater dated 21 March 1880). The rhyme starts: *"Dear Miss W., Sorry to trouble you, ..."*, and makes reference to *The Athenæum* and *The Examiner*. Spruce kept selected issues of these and other newspapers including the South American *Estrella do Amazonicas* and *La Union*, etc. (22 issues are at Manchester). In *The Reader* occurs a lengthy letter by Spruce on Beal-fires in the Amazon region,[7] and in *El Federalist* is an article about Richard Spruce.[8] Most of the papers and some other printed documents are dated during the 1880s.

The Spruce maps

The maps are of the Amazon region; 26 are printed, and 11 hand-drawn. Of the latter, some are apparently traced from printed maps, but others are clearly original, such as a rough sketch of the River Trombetas. This sketch may have preceded what is possibly the most interesting map of the collection, a finely detailed map of the same region with compass bearings and lines of latitude and longitude, drawn by Spruce from first principles. The caption explains how five points were fixed by astronomical observation and the remainder by compass bearings, and how he ascended the river in 1849.

Other maps of interest to those studying Spruce's journeys include a diagrammatic map of the Rio Uaupés, probably intended as a functional guide, like a modern London Underground map.

The Spruce mosses

Moving on from documentation to the actual specimens themselves, Manchester Museum has a very incomplete set of Spruce's *Musci Amazonici et Andini*. The specimens form a two volume hardback exsiccata, well bound and in good condition. Along the spine of each is printed in large gold letters "MUSCI AMAZONICI ET ANDINI", followed by "I" or "II"; across the base of the spine of volume I is neatly painted in white "1837 x", which presumably cannot be a date because it is too early, nor does it match any accession or registration number. There are no title pages or any other data that might indicate the origin of the volumes.

The specimens are stuck (some in packets) to individual pieces of paper, which are in turn mounted on the pages of the volumes. Also glued to the pieces of paper are small labels printed with "SPRUCE. *Musci Amazonici et Andini. No.* ", and filled in by hand (not Spruce's) with numbers, identifications, and (usually coded) locality. The localities are almost all very crude, e.g. "A.Q." (= Andes Quitensis), or "Fl.N." (= Flumen Negro).

The specimens correspond to those in Spruce's *Catalogus*;[9] the Manchester copy of this is marked "Muscorum Spruceanorum Catalogus; from the author", probably in the same hand that is found in the bound volumes, and is recorded also in the Herbarium Register in November 1895 but not in 1919. Also on the front of the Museum's copy of Spruce[10] is written in a different hand "2925; 3.9.95"; the four digit number is not an accession number. It is possible, then, that the exsiccata came to Manchester with the catalogue shortly after Spruce's death, possibly in 1895, as individual specimens which were then mounted and bound by the Museum.

There are in total 289 specimens in the exsiccata, numbered from 1 to 1515, but including 18 specimens sub-numbered "b", "c", etc. This leaves 1263 specimens missing from the main sequence which goes to 1518, plus 34 additional sub-numbers. The specimens are attached to the sheets in a sequence without any gaps; the missing specimens are not among the Spruce mosses incorporated into the main moss collections. There is now a complete list of the Manchester *Musci Amazonici et Andini* specimens, which include, for example, *Octodiceras hydropogon* (now *Fissidens*), and *Campylopus sprucei* (now *C. savannarum*).

Manchester's other Spruce mosses have been incorporated within the two main moss collections in the Herbarium: the British, and the General or non-British. These are large collections, and the distribution and percentage of Spruce material has been estimated by random dip-stick sampling. The Spruce mosses are (generally) stamped "SPRUCE HERB. PROP." but lack any accession numbers, and no record can be traced in any Museum register (except possibly as part of Kk854). Within the British collections, about 10% of the estimated 32,000 total is either collected by Spruce, or is from his personal herbarium although collected by others such as J.B. Wood; some specimens were collected after Spruce's death, for example by Slater in 1895. Within the General collections, probably no more than about 5% of the estimated 68,000 collections is Spruce material; much of this was collected by others such as Taylor, Gardner, Jameson and Purdis, and most of the material collected by Spruce himself is from the Pyrenees. A total of about 5,000 Spruce mosses is a reasonable estimate.

The Spruce liverworts, and the liverwort database

Spruce's liverworts were certainly his main interest, and it is undoubtedly his own personal and very large herbarium, including his British, Pyreneean, and South American collections, that we have at Manchester. Desmond 1994 mentions these plant collections.[11]

Spruce's personal herbarium came to Manchester Museum in two main lots. On 15 July 1919 we accessed: "Spruce Coll.n of West Indian Hepatics. (Mr Slater's) through W.H.P.", accession number Kk847; and on 29 November 1919 we accessed: "Herb. including Spruce's Bryophyta (few Ferns, Phanerogams) etc. European & General. Exor.s of Mr. Slater.", accession number Kk854. Other accession numbers indicate several other much earlier dates, for example some *Hepaticae Spruceanae* in 1891, 1892 (from Malton), and in 1894. These were purchased as sets.

A loose-leaf (but string-bound through punch-holes, with hard covers) 210 x 129 mm handwritten volume is entitled "List of Hepatics Spruce Collection Kk847 & Kk854". It is undated, but evidently post-March 1924. To the inside front cover is stuck the typewritten script: "In a letter at Kew Spruce states that he is sending a complete set of his Hepaticae Amazonicae et Andinae. He evidently did not retain a complete set in his own herbarium. There were many gaps in the set purchased from Slater's executors. Many of the specimens unrepresented in Spruce's own herb. are in purchased sets previously acquired by Manchester Museum ...", and the first page has handwritten: "Kk847 = Spruce Coll.n of W. Indian Hepatics purchased from Exor. of Mr Slater July 1919. In addition to these the Gen. Herb. (Hepaticae) of the MM contains a number of packets ex herb. Spruce, which were in the Pearson Coll.n of exotic hepatics Kk960 presented by Mrs Pearson March 1924. Also a purchased set K1412-K1843 Hepaticae Spruceana Amazonicae et Andinae acquired in 1891." The remaining 124 sheets list genera, and species prefixed by the number of packets of each, in the same handwriting. The volume has only recently been found (1995), and no further work has been done on it.

This volume, and the brief lines in the Herbarium Register are the sole information so far found in the acquisitions. Manchester Museum *Reports* were not produced from 1916-17 to 1922-23 (because of the Keeper's military service and "ill-health resulting therefrom"), and the Museum Committee minutes surprisingly make no mention of these massive acquisitions, whilst apparently recording in detail every other minor item. Matthew Slater was Spruce's executor. It is clear from a letter from Anthony Gepp (on headed BM(NH) paper) to W.H. Pearson dated 20 May 1919 that Spruce's own herbarium was in the hands of Slater who lent specimens out; Slater had in fact died in February 1919, although Gepp evidently did not know this.

The following data (revised since the original presentation of this paper) show the significance of the Spruce liverwort collections at the Manchester Herbarium (Herb. MANCH):

a	total plant collections	**2,000,000**
b	total liverworts	**34,346**
c	all Spruce liverworts[1]	**8,971**
d	Spruce liverworts designated TYPE[2]	**924**
e	later TYPE designations[3]	**200**
f	Spruce's personal herbarium accessed in 1919[4]	**8,264**

(1): including personal herbarium and *Hepaticae Spruceanae*

(2): including holotypes, isotypes, lectotypes, isotypes, n. sp., sp. n., etc.

(3): these were designated TYPE etc. since data were input to the database (see below), although there may be some overlap with (d)

(4): marked: Kk847; Kk854; and/or SPRUCE HERB. PROP.

The figures given are derived from our computer-held catalogue of the liverwort herbarium, using multiple search criteria (see Appendix B). In addition there are further specimens that, although not directly attributed to Spruce, bear evident connections, e.g. labelled in his handwriting, or bearing localities where only he is likely to have collected. Such specimens are not revealed by computer searches, nor are those where data have been incorrectly entered as in transcription errors such as "Spurce" for Spruce.

The Spruce lichens

Professor Brian Fox, who is currently working on our lichen collection, estimates that about 20% of our estimated 10,000 lichens are from Spruce's personal herbarium. This would total about a further 2,000 Spruce specimens. These are mostly stamped "SPRUCE HERB. PROP." but lack any accession numbers, and no record can be traced in any Museum register. Most of these are from the Pyrenees though some are collected in England, and others are gifts from other people such as J.D. Hooker. Some of these lichens are of considerable interest, but it is clear that Spruce did not apply the same attention to lichens as he did to liverworts.

A fine specimen of *Lobaria virens* bears collection data (as *Parmelia herbacea*, Bolton Woods, Wharfedale, Dec. 20th, 1841) in Spruce's hand, written on the reverse of a previously unused label printed: "SPRUCE – LICHENES AMAZONICI ET ANDINI, No. ". No other examples of this printed label have been found, and this one was presumably called into service by Spruce after his return from South America in June 1864, when repacketing earlier collections.

The Spruce vascular plants

It is clear from the Register that the Museum has a "few Ferns, Phanerogams" from Spruce's herbarium, accessed on 29 November 1919 (number Kk854). These numbers are so small that any attempt to estimate them by dip-stick sampling would be unreliable.

Summary

The large Spruce collections at Manchester Museum number over 16,500 items, estimated as:

a	his own personal herbarium of liverworts	8,264
b	additional liverworts such as distributed sets	700
c	his own personal herbarium of mosses	5,000
d	additional mosses in distributed set	289
e	his own personal herbarium of lichens	2,000
f	documentation such as letters and maps	300

ACKNOWLEDGEMENTS

I thank Professor Brian Fox both for his valued work on the Manchester Museum lichen collections, and for providing Appendix C of this paper. I thank Dr John Lowell for his valued work on rescuing the liverwort database (see Appendix B). I also thank Dr Anne Secord of the Cambridge Wellcome Unit for the History of Medicine, for information concerning Spruce's correspondence with other bryologists, and both Lisa Randall and Emma Telfer for their assistance in typing out Spruce's letters.

NOTES

(1) A.R. Wallace, in *Notes of a botanist on the Amazon and Andes* (R. Spruce, ed. A.R. Wallace), London, 1908.

(2) J.W. Franks, HERB. MANCH. A guide to the contents of the Herbarium of Manchester Museum, Manchester, 1973, (Manchester Museum publication NS.1.73).

(3) Wallace, op. cit. (1).

(4) H.M. McGill, 'The case of the missing journal', *Manchester Review* (1960-61), **9,** 124-8.

(5) Spruce, op. cit. (1).

(6) McGill, op. cit. (4).

(7) R. Spruce, 'Beal-fires', *The Reader* (1865), **6(151)** (Nov. 18), 569. [A long discourse on ritual, religious and social significance of beal-fires (bonfires), fire-worship, and acquisition of fire, in South America; about 1,880 words.]

(8) A. Ernst, 'El Doctor Ricardo Spruce', *El Federalista* (1867), **4(1,101)** (Apr. 13), 3. [An account of Spruce's South American travels and scientific explorations; about 1,080 words.]

(9) R. Spruce, *Catalogus muscorum fere omnium quos in Terris Amazonicis et Andinis, per annos 1849-1860, legit Ricardus Spruceus*, London, 1867.

(10) Ibid.

(11) R. Desmond, Dictionary of British & Irish botanists & horticulturists, &c. Rev. ed. London, 1994, 647-648.

(12) McGill, op. cit. (4).

APPENDIX A

Selected extracts from Spruce letters at Manchester

Notes on transcriptions

1) Spruce regularly abbreviates words by omitting part and terminating the word with the last letter(s) written superscript over a point (or perhaps sometimes a hyphen, or occasionally nothing at all); these are all transcribed with a point preceding the last letter(s) typed in normal size. On other occasions Spruce

abbreviates a word by substituting the last part with a point, and this distinction has been maintained.

2) Spruce almost invariably writes the digraph "ae" as the ligature "æ", as in "Hepaticæ", and this has been followed. Spruce also almost invariably writes "and" as a gamma-like ampersand (often slurred with the following word especially if a short one), and here a standard ampersand has been used.

Sample letter from Spruce to Matthew B. Slater

NOTE: The 15,000 moss specimens referred to in this letter, according to their description, do not seem to be part of those held at Manchester Museum.

25 March/92

Dear Sir

I am glad & thankful to learn you have completed *Frullania* & co. If you send them tomorrow in the big hamper, I have all *Plagiochila* & a few of the lesser genera ready to send by return on Tuesday. The spec.ns are all separated & ready to be placed in envelopes & labels attached, & I must try to get a good number of packets made to send shortly. The bulk of *Plagiochila* are Quitenian, but there are a fair number of them from the Amazon plain. I shall need a few more Andes Quitensis labels printed, but I will try to estimate the number before ordering.

I do not know what I sh.d do without your aid. In 1865 I was almost in despair about the Mosses. Mitten had had them in his hands for some years – I had been in England (mainly at Kew) a year – and still he seemed to get "no forrader". He told me he had selected good spec.ns of all for himself, but not half of them were named, & they had to be distributed with only no.s. I went down to Hurstpierpoint, & took lodgings, but I got very ill there and c.d rarely sit up to a table for anything – not even so much as I can now. I was fortunate in finding two bright girls whom I taught to fasten down the spec.ns, after I had placed them on the papers, & in the position, they were to occupy. It was no slight task, for I was very particular in having all peduncles, & all fallen calyptras, securely stuck down. Thus, working 3 or 4 hours a day, for 5 days a week, it took them a year & a quarter to stick down 15,000 spec.ns. They have both been long married, & have several children, but I still hear from them at long intervals, & the elder (married to a ship captain) has told me that her work on my mosses was the pleasantest job she ever had in her life, for she seemed to be continually making pictures.

I was glad to get a sight of Mitten's Memoir on Japan mosses at last, & I want you to come over & have some talk about it. He has found among Miss Hutchins' Bantry collections a new *Plagiochila*, which I think you and Mr Scully c.d refind if you w.d only try.

I am profiting by the mild weather to clear off a few matters at the microscope & with great heart.

Yours truly

RSpruce

Extracts from letters to Slater concerning politics

[29 Febr. 1880]

... I want you particularly to read the first article [Oct. *Edinburgh Review*] on "Germany since the Peace of Frankfurt". I have never been an admirer of that "brazen image" Bismark, which so many Englishmen are ever ready to fall down & worship; but it does one good to read such a thorough exposure of his crooked policy & its effects. On p.p. 309-312 you will find an instructive picture of what imperialism & a successful war have done for Germany in 9 years.

[21 Mch. 1880]

... England just now is a "house divided against itself", & the electioneering struggle promises to be very bitter. I fear there is little hope of a majority of Liberals in the new Parliament. Our party is terribly divided, & on minor matters too, which should be sunk if we w.d secure anything like unanimity – it is all very well to talk about "Measures not men" – but if I got a good man, like Bright or Burt, I would not mind his having a few crotchets of his own.

I fear the actual member for Malton is a snuff, & so is the one for Whitby; and altogether there are too few really energetic & working members in the Liberal ranks.

[12 Sept. 1880]

Why is the present Prime Minister like the reformer of Bryology?
Because he's Hed-wig!

Brilliant isn't it?

[Reference to the "Whig" party then in power, an old name effectively superseded after 1830 by "Liberal".]

[26 Octo. 1885]

... Do not bother yourself about the elections, for there is hardly a pin to choose between the two leading factions. If we c.d only have a Republic ...

[Wedn.y m.g; otherwise undated]

... I return W.W. Strickland's effusion. It is amusing and – ridiculous. It is very kind of the Stricklands, father and son, to take such pains to educate us new electors. We in the east of Yorkshire appear to be a shale, chalky lot, descendants of ammonites and sponges; in the west they have more of the millstone-grit in their composition. If we had displayed a little more grit we might not have had to throw up the sponge to the Tories.

Extracts from letters to Slater concerning *Fissidens*

NOTE: this correspondence is a sample of Spruce's interest in the difficult *Fissidens bryoides/viridulus/pusillus* group; similar extracts could have been selected for many other taxonomic problem groups, such as those around *Scapania undulata* (L.) Dum., *Lophocolea bidentata* (L.) Dum., or *Chiloscyphus polyanthos* (L.) Corda. These taxa are interesting because they still vex bryologists a hundred years on, and Spruce's words are as fresh and relevant now as they were then.

[13 April 1880]

... I sh.d like to see the doubtful *Fissidens pusillus*. The true plant may have nearly ripe fruit on Whit-Monday in Mowthorpe Dale & other of our woods. I w.d direct your special attention to the spring- & summer-fruiting *Fissidentes*; ...

[15 April, 1880]

... If the *Fissidens* be not *F. pusillus* I shall not know what to call it. It does not agree exactly with what I possess under that name, & still less with *F. bryoides*.

[18 April, 1880]

... I have gone over anew all the European species I possess, & I can now speak of them with some confidence. I cannot consider them very satisfactorily worked up in the new ed.n of Schimper's Synopsis. He does not follow Wilson in adopting the name "*Fissidens viridulus*", for he believes (perhaps correctly) that the original *Bryum viridulum* of Linnæus was what we have been used to call *Weisia controversa*. His "*F. incurvus*" is plainly an agglomeration of distinct <u>species</u> – if all the others in his book are to be accorded that rank – and includes not only the true "*incurvus*", but also Wilson's typical "*viridulus*", which I take to be truly distinct. I gathered the latter on Kirkham Hill (April, 1847), & it is undoubtably the Hackfall *Fissidens*. Did you ever gather *F. exilis* (= *F. bloxami* Wils.) on Kirkham Hill? It grows on lias clay at the foot of the hill by the road-side, & also in a ravine below Duffit's farmhouse.

Schimper's "*F. pusillus* Wils." is a most extraordinary plant, which, judging by its characters, I have never seen. Yet I have an original spec.n from Wilson under that name, & spec.ns enclosed by him of my own gathering quite agree with it... . You will perhaps have it under the name "*F. incurvus* var. *pusillus*", & "*F. bryoides* var. *pusillus*". It is perhaps our commonest *Fissidens* at Castle Howard. Pray examine the inflorescence in your spec.ns, & tell me how you find it: whether monoicous or dioicous, & the ♂ organs how placed. – Sch.r says the fl.s are sometimes <u>bisexual</u>, which I have never seen in our "*F. pusillus*".

[20 Dec. 1880]

... I have been wishing for a long time to write to Dr. Braithwaite. When I pluck up courage to do it, I will also lend him some of my *Fissidentes*. Before this wintry fit you might have found some in the park, but they do not bear frost well & are apt to get bleached in winter. 'Tis a group that wants thoroughly restudying, both in the field and the cabinet. The morning after the last no. of Journ. Bot. appeared, the post brought me Schimper's original specimens of the Pontefract "*Fissidens pusillus*" – sent from Kew without any request of mine. The inflorescence is certainly synoicous, & I shall have to call the plant *F. woodii* Mss. – Have you any spec.n of *F. inconstans*, found by Boswell near Oxford?

[31 Dec. 1880]

... Yesterday I received from Stabler a packet & cont.s, nearly all his own and Mrs Barnes's *Fissendentes* at least the smaller ones. Among Barnes's are some from Dr. Wood of the <u>synoicous *Fissidens*</u> gath.d at Pontefract & elsewhere, along with copious notes from Wood himself, from which it appears that this moss is Wood's type of *F. viridulus* Sw. & that he never thought of it as "*F. pusillus* Wils." We shall therefore have to study *F. viridulus* (& *incurvus*) all over again. I never noted any synoicous fl.s in *F. viridulus*, but then I do not suppose I ever sought for them. Pray examine the infl. of such spec.ns as you have under that name & let me know the result.

[23 Jan. 1881]

... In *Fissidens* you will note that Boswell sinks *F. inconstans* into *F. bryoides*. *F. viridulus* includes (I presume) "*F. pusillus* Schimp." but not "*F. pusillus* Wils.".

[28 Nov. 1885]

... You w.d not learn much from Mitten's paper on *Fissidens* (for the loan of which I thank you). He poses as a sort of Bryological pope – "such (he seems to say) are my decrees – I shall not condescend to explain them further than suits me, & it is for you to adopt & obey them".

[16 Jan. 1887]

Holt thinks your *Fiss. madidus* is only *F. pusillus*. But then what is *pusillus*?

[25 Jan/92]

... This m.g's post bro.t me the enclosed from Mrs Britton. You will see she wants the true *Fissidens incurvus* – with the hooked capsule.

Extracts from miscellaneous letters

[undated. Welburn. Friday night]

... The face of Nature will soon be sombre enough, without looking at it through glasses coloured to resemble London smoke. Will you therefore get your friend the Optician to send me a pair of perfectly colourless Eye-preservers, and pay him for them?

[12 Oct. 1883]

... I am not surprised that M.cAndrew was disappointed at people taking more notice of Puff-balls than of his Hepaticæ. I sh.d never think of making a show of such minute plants. To me, their collection & study has been its own exceeding great reward. ...

[2 Febr.y 1884]

... Hepaticæ are such picturesque objects – there is so much greater variety in their foliage & flowers than in mosses – that it is grievous to see what a fine opportunity was lost to improve on Hooker's work in Carr.n's British Hepaticæ.

[26 Febr. 1884]

... I shall be glad to examine the monoicous *Lophocolea*. The scraps I retained are uni-sexual – ♂ plants closely entwined with the ♀, but distinct. It is only by observation of each individual case that the value of characters derived from the inflorescence can be estimated.

[Sat. m.g: otherwise undated]

... I am returning your hamper & in it you will find 3 parcels, containing the first 3 sets of my Exsiccata. They are all properly addressed & stamped, & I shall be obliged to you to post those addressed to Kew & Edinburgh, but to retain the one for Carruthers at the Brit. Museum until further notice. It might indeed be as well to keep the two first at your place until Monday, for if sent off today they c.d not be delivered tomorrow. This I leave to your better judgement.

The very day you called here with Barnes I had a letter from Geo. Moore of the Brit. Mus. to say that he had a set of Hepaticæ of the British W. Indies for me & w.d send it in a day or two. I have however not yet received it, & they are such a slippery set at the Museum that I do not propose to send their set of my hepat.s until I receive that parcel.

[undated. Coneysthorpe]

... If not too much trouble will you get me <u>a stone of best Small-grained Sago</u>, at Sewell's & <u>pay for it</u>. I have not had it good lately, & know not where to go for it. What I am using now seems mostly potato-starch with a very little admixture of some Sago.

APPENDIX B

Manchester Museum Liverwort Database

Manchester Museum's project to enter data from the Botany collections onto computer ran from 1980 to 1984, under the government-funded Job Creation Scheme and the Manpower Services Commission. Data capture and data input for the liverwort collections alone took about ten man-years, in addition to the time involved in supervising, the disruption of loans and other routine functions, and cost of hardware, plus additional overheads.

The database had only been available for editing and manipulation for about three weeks before the Spruce Conference in September 1993; improved search criteria applied since then have revealed over double the number of Spruce specimens mentioned at that time. The delay of 11 years before effective access to the information was almost entirely due to problems in transferring the FAMULUS database from the University mainframe (where it could not in practice be edited), onto Departmental PCs in a dBase file structure and using a suitable database management system (currently PC-File 7.0).

We are particularly grateful to Dr John Lowell, whose special botanical and computer skills have recently provided us with much improved global sub-string editing facilities; these he has already used to weed and correct errors in the current name field. Although much work still needs to be done to continue correcting the database, we now have a useful tool for those who wish to research our Spruce (and other) collections.

Two typical edited records are shown below:

Example 1

CURRENTNAM	Strepsilejeunea involuta (Gott.) Steph.
SYNONYMS	Lejeunea involuta Gott.; Crossotolejeunea torticuspis Spruce
LOCALITY	Amazon forest. Rio Negro et Taruma. San Gabriel; Habt: ad arbores
COLLECTOR	Spruce, R.
STATUS	ISOSYNTYPE, ISOLECTOTYPE
GCOD	AM5
PREVIOUS_N	MM KK854
ASSOC_NAME	Spruce, R. (Hepaticae Spruceanae Amazonicae)
ASSOC_DOCS	ISOSYNTYPE, ISOLECTOTYPE, Rev. GROLLE R. 1978
NOTES	Size 3 cm
ASSOC_SPP	
GENUS_NO	999
BIOL_STATE	
COMPUTER_N	21417

Example 2

CURRENTNAM	Mastigolejeunea innovans (Spruce) Steph.
SYNONYMS	Lejeunea innovans Spruce
LOCALITY	Amazon forests, Panure fl. Uaupes; Habt: in cortice ramulisque
COLLECTOR	Spruce, R. (CD: 1852)
STATUS	HOLOTYPE
GCOD	AM5
PREVIOUS_N	MM KK854
ASSOC_NAME	Spruce, R. (Herb. Prop.); Spruce, R. (Hepaticae Spruceanae Amazonicae)
ASSOC_DOCS	HOLOTYPE of Lejeunea innovans: det. GRADSTEIN 1991
NOTES	Size 3 cm
ASSOC_SPP	
GENUS_NO	130
BIOL_STATE	Fruiting
COMPUTER_NO	15895

APPENDIX C

Spruce in Manchester Central Library

Brian W. Fox

Tryfan, Longlands Road, New Mills, via Stockport, Cheshire SK12 3BL

Like many members of the excellent Richard Spruce meeting in York, in September 1993, I was made aware of a remarkable man that I had little knowledge of before. Following the Spruce Symposium I visited the Central Library in Manchester, where I had been advised by Professor Richard Schultes that several items referring to Richard Spruce were housed in the Central Library. There are three well archived and preserved collections of Spruce material:

1: **MSF 925.8 SP1**. A bound volume of 61 items of correspondence between Richard Spruce and others, such as the India Office, Sir Clements Markham, Royal Geographical Society, and the Linnean Society. A copy of his register of Baptism, his actual Passport signed by the Earl of Malmesbury, the Scroll presented to Spruce as Doctorem Philosophiae from the Academie Germanicae Naturae Curiosorum, a document from the Societas pro Fauna et Flora (Helsingfors) naming Spruce as a member, his Membros Honorarius from the Sociedad de Caracas, a document from the Botanical Society of Edinburgh (1886) on his election as member, and a letter from the Linnean Society on election to associate membership at the age of 76 as well as an interesting pencilled draft of his reply. In addition there are a series of pencilled lists of illustrations, several ink drawings of a ruin, rocks at Mowthorpe Dale, as well as a study of three types of leaves which appear to have been made in 1834 (aged 17!). Finally there is a number of listed receipts from Carlos Nash and others, dated around 1860.

2:   **MSF 581.98 SP1**. A bound volume of the early part of his manuscript journal from 7 June 1849 to 6 January 1850 of *Notes of a botanist*. The introduction indicates that the notes covered the period of 1849–1857; there thus appear to be seven years missing. The journal is a hand-written document of foolscap size.

3:   **MSF 925 WA1**. A volume of letters from Alfred Russel Wallace to Matthew B. Slater recording his efforts over several years trying to locate the missing Spruce journals, cajoling and remonstrating with Slater to search and find the documents. Wallace even employed a clairvoyant to try to discover the whereabouts of the missing papers and a description of a house and a cellar was given. Without a lock of Spruce's hair however, the experiment was apparently not as successful as it might have been. Attempts to match this description with the residences of a Mr Teasdale as well as others apparently failed. One letter includes the frustrated sentence *"cannot you get a strong woman to help your daughter make a <u>thorough search</u> from <u>attic</u> to <u>cellar</u> in <u>cupboards, boxes & bundles, & everywhere else</u>, till they are found"* (6 September, 1906). An account of these letters was detailed by McGill.[12]

The last letter (22 February, 1907) refers to the Castle Howard family and asks Mr Slater if he thought that they would contribute to the publication of some native drawings of Quito costumes.

The first letter from Wallace to Slater (31 December, 1893) includes the comment *"to you who saw him so frequently, the loss must be a great one, since I hardly know anyone who was so intelligent, amusing and lovable a companion"*.

23

The Spruce collections in the Herbarium of Trinity College, Dublin

John A. N. Parnell

The Spruce collections in the Herbarium of Trinity College, Dublin

John A. N. Parnell

School of Botany, Trinity College, Dublin, Ireland

Background

The Trinity College Herbarium (TCD) was established in 1840; its nucleus was the personal collection of Thomas Coulter, the first curator. However, the herbarium is really the creation of William Henry Harvey who was curator from 1844 until his death in 1866. Harvey's own hard work combined happily with an ability to make and keep friends (for example the herbarium still contains specimens sent to him by Charles Darwin from his voyage in *The Beagle*). This, together with his own extensive exploration of Australia, North America and South Africa led to the accumulation of 100,000 specimens. Although his herbarium budget was small, Harvey managed to purchase a number of collections which have now become classic. Among these were collections made by Richard Spruce.

The Spruce collections

In all, we have about 3,200 vascular plant specimens collected by Spruce.[1] These were mostly collected in Brazil (mainly from the Amazon Basin); but there are also quite a few from the Andes of Ecuador (850), rather fewer from Peru and Venezuela and very few from Colombia; all collections date from between 1849 and 1859 with the commonest collection numbers between 2000 and 4000. The geographical pattern of collection in the pteridophyte collection[2] appears to be different from the higher plant collections. We have about 505 pteridophytes collected by Spruce; 124 of these come from the Andes of Ecuador, 115 from Peru Orientalis, 65 from Pará province and about 40 from the Rio Negro. Among those, the most frequently encountered collection numbers are between 3000 and 5000. Finally, in addition to our higher plant material, we have some bryophytes collected by Spruce before he went to South America. These were gathered between 1845 and 1846 from the Pyrenees and S.W. France. There are no lichens collected by Spruce in the herbarium.[3]

Clearly, one interpretation of the difference in geographical breakdown of our pteridophyte and higher plant collections is that Harvey purchased them differentially, concentrating on the Andes for one and the Amazon Basin for the other. However, the recent article by Henderson *et al.*[4] implies that the Andes are botanically exceptionally rich; and indeed they are richer in species of pteridophyte than the Amazon Basin. So the fact that our collection of Spruce material contains a greater number of pteridophytes from the Andes than from the Amazon Basin and a greater number of flowering plant specimens from the Amazon Basin than from the Andes may simply represent biological reality.

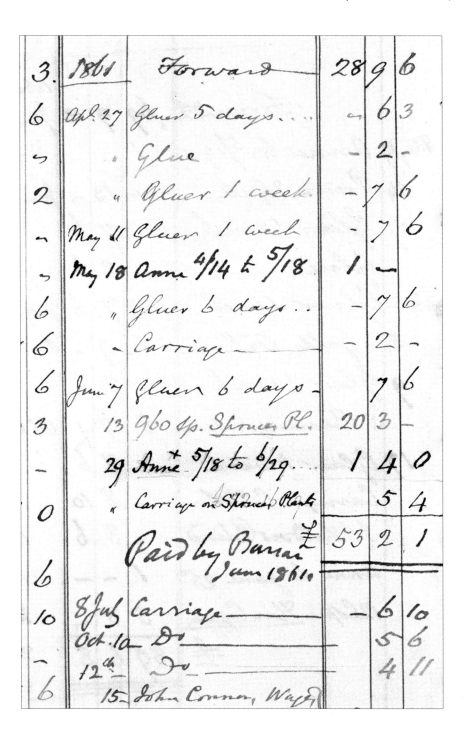

Fig. 1. Harvey's manuscript accounts for the Herbarium of Trinity College Dublin showing for June 13th 1861 the purchase of 960 Spruce specimens; followed on June 29th by payment of carriage for the same of 5/4d; also note the various payments (e.g. 6/3d on April 27th 1861; made to the Gluer at the rate of 1/3d per day.

Unfortunately our herbarium records for the period of Harvey's curatorship are fragmentary but surprising details still lurk and we can trace the purchase of most of the Spruce material in the herbarium (Fig. 1, Table 1).

Harvey clearly recognised the significance of the Spruce material as he spent a large portion of his budget on it: for example, the purchase of 5th July 1858 took over half of his total annual budget for specimen purchase (= £30).[5] In total the Spruce specimens cost Harvey about £66. It is difficult to grasp what this figure means in present-day terms. However, it may be helpful to recall that in 1856-1857, in Dublin, the daily wage of the herbarium's gluer was 15 old pence per day (Fig. 1), that of an unskilled labourer 18-24 old pence per day and that of a painter 56 old pence per day.[6] This equates with maximum yearly wages of about £19, £29 and £73 respectively (assuming no Sunday working). Indeed, Spruce's annual government pension was only £50 in 1865.

We have data (Table 1) showing that Harvey purchased 2759 specimens for a total of £57.19s.0d or about 5d per specimen. This means that the £8.7s.0d for which no details are available (Table 1) probably bought about 400 specimens. In all, then, we know that Harvey bought approximately 3159 specimens: this is remarkably close to Webb's estimate,[7] based on random sampling, of 3,200 Spruce specimens in the collection. This seems to indicate that very few of the Spruce specimens which Harvey bought have been lost, sold or sent out for exchange. However, Spruce collected at least 7,000 specimens.[8] Why then did Harvey purchase only 3,200 of them; why did he begin purchasing in 1857 (eight years after Spruce reached Pará) and stop in 1861 (when Spruce was at Duale)? Unfortunately I have been unable to come up with any rational reasons for this; but as many TCD specimens have numbers in the 4000-5000 range it may be that Harvey, who began his purchases relatively late, could get no more.

Nearly all of the Spruce specimens in TCD are incorporated into the main collection, though a few particularly puzzling ones still remain to be placed in the appropriate generic folder. A much larger number still await confirmation of their specific identity; these latter are particularly worrying as many of them are labelled "N.Sp." (i.e. new species) by Spruce. Theoretically, as most of our Spruce specimens are numbered it would be possible, given sufficient time and manpower, to trace

Table 1. Details of specimens collected by Spruce and purchased for the TCD herbarium by W.H. Harvey.

DATE	NUMBER OF SPECIMENS	PROVENANCE	PRICE	CARRIAGE
30th June 1857	695	South American Plants	£14.13s.0d	Unknown
5th July 1858	770	Unknown	£16.3s.0d	Unknown
10th July 1859	Unknown	Unknown	£8.7s.0d	3/6d
25th August 1860	334	Unknown	£7.0s.0d	Unknown
13th June 1861	960	Unknown	£20.3s.0d	5/4d

Fig. 2. Isotype of *Licania intrapetiolaris* Spruce ex Hook.f. showing, at the foot on the left-hand side, a typical label of the Spruce collections in TCD and, opposite, a determinavit slip of G. Prance.

the appropriate name through the literature. This is not as simple a procedure as it might seem, because a significant proportion (perhaps 15%) of TCD's Spruce material is without a collection number at all and a very small number of specimens bear two collection numbers. In the latter case both numbers appear to be original and there is no systematic difference between the pairs on different sheets (numbers may differ from each other by more than 1,300 or less than 120). As some of the sheets with two numbers also bear two or more plant fragments it may be that a single label was used for two separate collections of the same taxon from the same locality.

The current physical condition of the Spruce material is generally excellent as its rate of use has been low over the past century (Fig. 2). Recent examination of certain parts of the collection *e.g.* the Lecythidaceae, Chrysobalanaceae, Apocynaceae and Aquifoliaceae have shown that there is a large amount of type material present. As most of our Spruce material is in much better condition than corresponding material held in larger institutions, workers on the tropical South American flora would be well advised to look at our collection before nominating lectotypes.

ACKNOWLEDGEMENTS

I thank Mr B. Dempsey, the College Photographer, for taking the included plates.

NOTES

(1)   D.A. Webb, 'The herbarium of Trinity College, Dublin: its history and contents', *Botanical Journal of the Linnean Society,* (1991), **106**, 295-327.

(2)   J. Parnell, 'The Pteridophyte herbarium of Trinity College Dublin', *Fern Gazette* (1985), **13**, 47-48.

(3)   J. Parnell, 'The lichen collection of Trinity College Dublin', *Lichenologist* (1982), **14**, 291-292.

(4)   A. Henderson, S.P. Churchill and J.L. Luteyn, 'Neotropical plant diversity', *Nature* (1991), **351**, 21-22.

(5)   E.P. Wright, 'The Herbarium of Trinity College Dublin: a retrospect', *Notes from the Botanical School of Trinity College, Dublin* (1896), **1**, 1-14.

(6)   F.A. D'Arcy, 'Wages of labourers in the Dublin building industry 1667-1918' *Saothar* (1989), **14**, 17-32.; F.A. D'Arcy, 'Wages of skilled workers in the Dublin building industry 1667-1918', *Saothar* (1990), **15**, 21-37.

(7)   Webb, op. cit. (1).

(8)   R. Spruce, *Notes of a botanist on the Amazon and Andes* (ed. A.R. Wallace), London, 1908, 2 vols.

24

Richard Spruce specimens in the Ulster Museum Herbarium, Belfast, Northern Ireland

Catherine R. Hackney

Richard Spruce specimens in the Ulster Museum Herbarium, Belfast, Northern Ireland

Catherine R. Hackney

Dept. of Botany, Ulster Museum, Botanic Gardens, Belfast, NI

The Ulster Museum Herbarium (BEL) is an amalgamation of the herbaria of the Belfast Museum and Art Gallery and the Queen's University of Belfast. The combined herbaria currently hold an estimated 100,000 specimens, mostly from the British Isles and Europe, and include Spruce specimens. The oldest Museum specimens date from the 1790s, while the University collections were apparently started shortly after the foundation of the University itself in 1845, and contain no specimens from before that date. No records were transferred with the University herbarium, so the means of acquisition of their Spruce material is unknown.

There are approximately 200 Spruce specimens in the Museum's vascular plant herbarium. They were collected between July 1849 and August 1852, and are mostly from the Amazon basin around the Rio Negro, with a few from the watershed area between the Rio Negro and the Orinoco. About 30 genera are

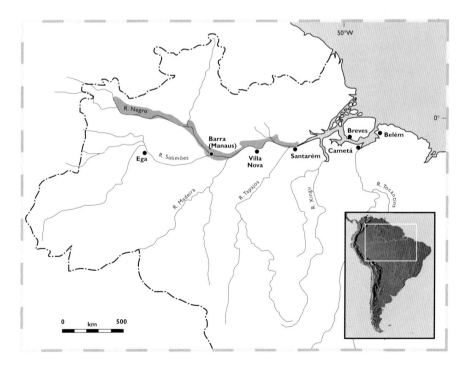

Fig. 1. Areas from which the Ulster Museum specimens were collected in 1849-1852.

represented, with the largest numbers being from the Rubiaceae, Myrtaceae, Melastomataceae and Chrysobalanaceae. A full species list is given in the Appendix, with species names quoted as they appear on the herbarium labels; in this list no nomenclatural changes have been made.

Nothing is known of the history of the Spruce specimens. It had been thought that they might have been duplicates exchanged with the herbarium in Trinity College, Dublin, but this now appears unlikely, as there is material in BEL which is not in TCD, and nearly all Spruce material acquired by TCD can be accounted for. As much of the Ulster Museum material is identical to the TCD specimens, it can only be assumed that the BEL material is part of one of the sets of duplicates prepared by Bentham for subscribers in the last century.

As with TCD specimens (see Chapter 27, this volume), those in BEL are generally in very good condition, having been little used since their acquisition by the University; it is also noteworthy that at least 6 specimens are identical in all respects to specimens nominated as type material in TCD. Although the collection is small, it might be advisable for researchers to consult BEL specimens when working further on Spruce material.

REFERENCES CONSULTED

A. Gepp, 'In memory of Richard Spruce', *Journal of Botany* (1894), **32,** 50-53.

R. Spruce, *Notes of a botanist in the Amazon and the Andes* (ed. A.R. Wallace), London, 1908, 2 vols.

G. Stabler, 'Obituary notice of Richard Spruce, Ph.D.', *Transactions and Proceedings of the Botanical Society of Edinburgh* (1894), **20**, 99-109.

D.A. Webb, 'The Herbarium of Trinity College, Dublin: its history and contents', *Botanical Journal of the Linnean Society* (1991), **106,** 295-327.

APPENDIX

Richard Spruce specimens in the Ulster Museum (BEL)

Species List

FAMILY	SPECIES	SPRUCE LABEL NO.
Ochnaceae	Blastemanthus grandiflorus Sp. n.	8018
	Sauvagesia salzmannii Benth.	-
Caryocaraceae	Caryocar glabrum Pers.	1872
	C. villosum	1819
Guttiferae	Calophyllum sp.	1908
	Clusia sp.	2251
	Clusia sp.	2062

Guttiferae	C. insignis Mart.	1805
	Caraipa paniculata	1915
	Garcinia sp.	2377
	Tovomita amazonica Popp.	1492
	Tovomita aff. T. umbellata Benth.	1513
	Vismia cayennensis Pers.	-
	V. dealbata HBK [Kunth]	-
	V. guianensis Pers.	2170
	Vismia Sp. n. aff. macrophyllus	-
	V. obtusa Sp. n.	-
Flacourtiaceae	Homalium pedicellatum Sp. n.	1489
Turneraceae	Turnera opifera Mart.	-
Cucurbitaceae	Anguiria Sp. n.	1493
Sterculiaceae	Sterculia aff. S. frondosa Kuli	1685
	S. striata P. Hut. & Vaud.	-
Malvaceae	Sida glomerata Cav.	-
	Urena americana L. f.	2255
Euphorbiaceae	Pogonophora schomburgkiana Myers	2083
Chrysobalanaceae	Couepia bracteosa Benth.	1496
	C. bracteosa Benth. var. interrupta	2003
	C. eriantha Sp. n.	-
	C. leptostachya Sp. n.	1536
	C. myrtifolia Sp. n.	2262
	Hirtella americana Aubl.	2367
	H. rubra Benth.	-
	H. physophora Mart. & Zucc.	-
	Licania costata Sp. n.	2197
	L. floribunda Benth.	1831
	L. floribunda Benth. var. acuminata	1830
	L. heteromorpha Benth. var. pilifera	1649
	L. incana Aubl.	1953
	L. miquilioides Sp. n.	1801
	L. myristicoides Sp. n.	-
	L. obovata Sp. n.	1569
	L. pallida Sp. n.	1576?
	L. ramoupessia Sp. n.	-
	L. triandra Ch. & Schl.	-
	Parinarum brachystrychium Benth.	1577
Connaraceae	Connarus Sp. n.	2061
	C. ruber Planch.	-
Combretaceae	Combretum variabile Mart.	-
	Terminalia suaveolens Spruce Sp. n.	1887

Lecythidaceae	Gustavia brasiliana DC.	-
	Lecythis sp.	1797
	L. aff. ovalifolia Mart.	-
	L. ovalifolia Mart. var?	1884
	L. pachysepala Spruce	1912
	L. retusa Spruce Sp. n.	-
Myrtaceae	Calyptranthes Sp. n.?	1815
	C. cuspidata Mart.	1905
	Campomanesia myrtogoethiana Mart. aff.	1804
	Eugenia sp.	1663
	Eugenia sp.	2774
	Eugenia sp. (4) ... humilis	-
	Eugenia sp. (5)	-
	Eugenia sp. (6)	-
	Eugenia sp. (7)	-
	Eugenia sp. (9)	-
	Eugenia sp. (10)	-
	Eugenia sp. (11)	-
	Eugenia sp. (12)	1458
	Eugenia sp. (13)	-
	Myrcia sp.	-
	Myrcia sp.	1828
	Myrcia sp.	1904
	Myrcia sp. (2)	-
	Myrcia sp. (6)	-
	Myrcia sp. (8) var.	-
	Myrcia sp. (11)	-
	Myrcia sp. (15)	-
	Myrcia sp. (16)	-
Melastomataceae	Bellucia circumcissa Spruce	2162
	Clidemia / Oxymeris	2263
	Clidemia (3) / Staphidium	-
	Clidemia (5) / Staphidium	-
	(Staphidium sp.)	2103
	(S. spicatum)	2020
	Loreya spruceana Benth.	-
	Maieta guianensis Aubl.	2163
	Miconia sp. / Diplochita mucronata DC.	-
	Miconia sp. (3)	-
	Miconia sp.	-
	Miconia sp.	2166
	Miconia sp.	1878
	M. impetioloris aff.	2134
	M. lepidota DC.?	-
	M. tomentosa DC. var.	-
	M. tomentosa DC.	2335
	Myrmidone macrosperma Mart.	2336
	Pleroma tobichouna aff.	-

Melastomataceae	Rhynchanthera monodynama DC.	-
	Spennera dysophylla Benth.	-
Memecyclaceae	Mourira apiranga Spruce	-
	M. brevipes Hook.	1935
	M. eugenaiefolia Spruce	-
Onagraceae	Jussiaea octonervia Lam.	2413
Anacardiaceae	Anacardium giganteum Hanc.	1971
	A. spruceanum Sp. n.	1684
	Thyrsodium schomburgkianum Benth.	1749
Burseraceae	Icica pubescens Spruce	-
	I. unifoliata Spruce Sp. n.	1960
Sapindaceae	Castella paullinioides Spruce Gen. nov.	2169
	Cupania laxiflora Benth.	-
	C. spruceana Benth.	-
	Schmidelia amazonica Mart.	1596
	Serjania nitida Benth.	-
Houmiriaceae	Sacoglottis amazonica Mart.	-
Erythroxylaceae	Erythroxylon sp.	1854
Malpighiaceae	Burdachia macrocarpa Sp. n.	-
	Byrsonima crassifolia Kunth. var.	1919
	B. inundata Benth.	1911
	Byrsonima, pet roseis, B. nitidifolia aff.	1628
	Byrsonima Sp. n. aff. B. peruviana	2073
	B. spicata Rich. var. angustifolia	-
	Pterandra latifolia A. Juss.	-
Vochysiaceae	Salvertia convallariaeodora A. St. Hil.	-
Polygalaceae	Securidada bialata Sp. n.	-
Araliaceae	Hedera resinosa Sp. n. var. brevifolia	2350
	H. resinosa var. ramosissima	2349
Vitidaceae	Cissus aizoides L. var.	-
Loranthaceae	Loranthus sp. (1)	-
	Loranthus sp. (6)	-
	Loranthus sp. (8)	-
	Loranthus sp. (13)	-
	Loranthus sp. (14)	-
	Loranthus sp. (15)	-
	Loranthus sp. (17)	-
	Loranthus sp. (20)	-
	Loranthus sp. (21)	1550
Viscaceae	Viscum sp. (4)	-

Rubiaceae	Amaioua petiolaris Sp. n.	2177
	Bertiera leiantha Sp. n.	1788
	Borreria parviflora Mey.	2374
	B. scabisoides Ch. & Schl.	-
	Commianthus concolor Spruce Sp. n.	2028
	Cordiera myrciaefolia Sp. n.	-
	Eukylista spruceana Gen. nov.	-
	Faramea longifolia Benth.	?2025
	F. vaginata Sp. n.	-
	Genipa cymosa Sp. n.	-
	G. nervosa Sp. n.	-
	Geophila arenaria Spruce Sp. n.	2285
	Gomphosia laxiflora Benth.	1707
		1952
	Ixora spruceana Benth.	-
	Oldenlandia herbacea DC.	2372
	Palicourea sp. (1)	-
	Palicourea sp. (2)	-
	P. pallidiflora Spruce	2137
	P. riparia Benth.	1929
	Patabea ornata Spruce	-
	Perama hirsuta Aubl.	-
	Psychotria barbiflora DC.	-
	P. bracteata DC.	1847
	P. bracteata DC. var. ?	1894
	P. limbata Sp. n.	1902
	P. limbata Sp. n.	1655
	P. nervosa Benth.	1895
	P. nervosa Benth. var. ?	1696
	P. subundulata Benth.	2018
	Remijia tenuiflora Benth.	2079
	Rudgea longistipula Sp. n.	2301
	R. villosa Sp. n.	1277
	Schradera spicata Sp. n.	-
	Sipanea hispida Sp. n. aff. S. dichotoma	2051
	S. trichantha Miq. / S. dichotoma var.?	-
	Sphinctanthus maculatus Spruce	1939
	S. rupestris Benth.	1874
	Sprucea rubescens Benth. Gen. nov.	1601
Compositae	Acanthospermum xanthoides DC.	-
	Bidens bipinnata L.	-
	Clibadium asperum DC.	2136
	Eupatorium Sp. n. aff. E. ferrugineum Gardn.	-
	Mikania sp.	1585
	M. polystachya DC.	2304
	M. scandens L. var.	-
	Pectis elongata HBK [Kunth.]	-
	Trichospora menthoides HBK [Kunth.]	-

25

Spruce material at the Royal Botanic Garden, Edinburgh

Jennifer Lamond

Spruce material at the Royal Botanic Garden, Edinburgh

Jennifer Lamond

Royal Botanic Garden, Edinburgh, UK

In the herbarium of the Royal Botanic Garden, Edinburgh (E), there is a large quantity of phanerogams and cryptogams collected in South America by Richard Spruce in the 1850s, together with some bryophytes from the Pyrenees (1845-46) and a few from Britain, e.g. Galloway, Teesdale c. 1843. There are also some letters from or concerning Spruce in the library archive.

HERBARIUM MATERIAL

Flowering plants and ferns: S. American

About 3600 were acquired between 1856 and 1861, presumably purchased: they were catalogued as each batch arrived (eight in all), some in systematic order, mostly named. A further 400 were added with the permanent loan of the Glasgow University's phanerogamic herbarium in 1965 (E-GL), which is partially listed on cards – the monocotyledons are completed, while the dicotyledons have been done as far as the Compositae, in the Bentham & Hooker (B&H) system; the cryptogams are still at Glasgow (GL). Ferns were also acquired in other collections, e.g. from Hooker's herbarium; Rothery's plants; & Neill Fraser's herbarium. Many of these specimens are types.

Cryptogams

Cryptogams 'Amazonici et Andini' are listed as being purchased in 1866; most are unnamed: they comprise 110 lichens, 19 algae, 71 fungi and 1050 mosses. There are also 37 'seaweeds and zoophytes from shores of the Pacific'.

Hepatics were added later, according to correspondence at RBG Edinburgh.

SUMMARY OF SPRUCE CORRESPONDENCE AT RBG EDINBURGH

Letters from Spruce

A. to John Hutton Balfour; [Queen's Botanist, 1845–79].

12 May 1844 from Collegiate School, York. [There had been previous letters not now extant]: comments on East Indian mosses determined by 'Wilson' and by 'Taylor': also on shells and J. Backhouse.

B. to P. Neill Fraser

31 July 1892 from Coneysthorpe, near Malton, Yorkshire: On South American hepatics – examined several collections – regrets publishing, crudely without proofs, in *Bulletin de la Société Botanique de France* – offers set of own collections for RBG Edinburgh; manuscript for *Botanical Society Transactions?* (will be sent via Slater).

C. to Isaac Bayley Balfour [Professor of Botany & Regius Keeper, RBG Edinburgh 1888-1922]

24 August 1892 from Coneysthorpe, near Malton, Yorkshire: Returning loaned specimen and gifting other; offers South American hepatic set for Edinburgh, for 30 shillings per 100, best sets containing about 400 specimens; & on paper on new hepatics for *Botanical Society Transactions*.

Letters concerning Spruce

A. G. Stabler to Isaac Bayley Balfour, 3 February 1894 summarises contents of a manuscript he is to send IBB on Spruce – (contains humorous anecdotes).

B. P. Neill Fraser to or from Stabler, Slater, Terras, I.B. Balfour March-April 1896 – all to do with hepatic manuscript apparently mislaid; Slater's contains notes on what Spruce was working on in his last few years.

26

Archival resources on Richard Spruce: a preliminary listing

Sylvia FitzGerald

Archival resources on Richard Spruce: a preliminary listing

Sylvia FitzGerald

Royal Botanic Gardens, Kew, Richmond, Surrey, UK

Field notes, lists of excursions, plant lists etc.

List of plants mainly from near Ganthorpe. ms. [?1834]. [Formerly in the possession of G. Stabler. *Present location unknown.*]

List of the flora of the Malton district. Unpublished ms. [?1837]. [Formerly in the possession of M.B. Slater; see Sheppard (1909) p. 46. *Present location unknown.*]

List of botanical excursions, June 19, 1841-April 30, 1863. *At the Linnean Society of London.*

Manuscript field journal, 7 June 1849-10 Dec. 1850. With another 'fair copy' of the Journal, 7 June 1849-6 Jan. 1850. *At Manchester Central Library (Archives Dept.)* MSF 581.98 SP1.

Meteorological registers, Liverpool to Para, 1849, Rio Negro and in Quitenian Andes, 1851-62; notes for description of Tarapoto, eastern Peru; miscellaneous notes on his travels in South America etc. *At the Royal Geographical Society, London.*

Journals from Barra to the Orinoko, from Barra to Tarapoto, and notes on the uses of Amazon plants and on cryptograms [Plantae amazoniae. Domestic uses pp. 1-30; Stirpes cryptogamia sp. 1-279]. 1851-1855. *At the Royal Botanic Gardens, Kew, Library & Archives, with the Spruce Papers.*

Notes on the S. American Mosses in the herbarium of Dr Thos. Taylor, 1847. Notes on mosses in the herbarium of Linnaeus and of Sir J.E. Smith. Notes on the mosses of his own herbarium, 421-790. Journal... Tarapoto in Peru lat. $6^1/_2$°S to Baños in Ecuador...1857. *At the Royal Botanic Gardens, Kew, Library & Archives, with the Spruce Papers.*

Plantae amazonicae, 267-3846. 1849-1855. 2v. *At the Royal Botanic Gardens, Kew, Library & Archives with the Spruce Papers.* (There is no record of nos. 1-266, other than specimens in Bentham's herbarium, now at Kew.)

Plantae amazonicae. Domestic uses pp. 31-61; and miscellaneous notes (including ethnographic notes). 1855? *At the Royal Botanic Gardens, Kew, Library & Archives, with the Spruce Papers.*

Plantae andinae, 3851-6551, 6576-6580. 1855-1857. 2 vols. *At the Royal Botanic Gardens, Kew, Library & Archives, with the Spruce Papers.*

Spruce's Ecuador ferns. 242 species Feb. 12th/51. In: RBG Kew. Plant [determination] lists, Vol. 2. Orient. Africa. America. 1852-67, fol. 88-91, ms. *At the Royal Botanic Gardens, Kew, Library & Archives.*

Spruce's plants from redbark woods of Chimbaraze 1861 (rec'd Aug.). In: RBG Kew. Plant [determination] lists, Vol. 2, Orient. Africa. America. 1852-67, fol. 88-91. *At the Royal Botanic Gardens, Kew, Library & Archives.*

Letters by Richard Spruce

[Letters]. Correspondence etc. and Cinchona, 1845-1880. *At the Royal Botanic Gardens, Kew, Library & Archives, with the Spruce Papers.*
> Letters to George Bentham 1845-78; notes on collections; letters to Sir Joseph Hooker 1864-80; printed report on Cinchona 1860-62; poster announcing award of Doctorem Philosophae... Dresden, 1864. A further letter is in Bentham Correspondence Vol. 9 (3701).

[102 letters to Sir William Hooker & Sir Joseph Hooker 1839-1882]. *At the Royal Botanic Gardens, Kew, Library & Archives, in Directors' Correspondence.*

Letters to J.G. Baker doc. 81 (6 April 1869). *At the Royal Botanic Gardens, Kew, Library & Archives.*

Letters to William J. Borrer, 1842-1848, and miscellaneous notes, some in ms. by Lindsay Fleming. *At Royal Botanic Gardens, Kew, Library & Archives, with the Spruce Papers.*
> Also: a ms. copy of these letters, transcribed by Lindsay Fleming, and enlarged photostat copies of the originals, with the Borrer Papers (1960s).

William Munro correspondence. doc. 196 (4 April 1849). *At the Royal Botanic Gardens, Kew, Library & Archives*

Letters from Richard Spruce, 1856-75, together with some letters from Hanbury to Spruce presented by Alfred R. Wallace in 1908, and correspondence relating to the gift. *At the Royal Pharmaceutical Society of Great Britain, London, with Daniel Hanbury's Collections of correspondence on materia medica and related subjects.*

Letters [to Matthew B. Slater, 1880-1893, & others, with miscellaneous documents, & 22 newspaper extracts], c. 1880s. *At the Manchester Museum.*

[Letters to J.H. Balfour, P.N. Fraser, I.B. Balfour]. *At the Royal Botanic Garden, Edinburgh.*

Letters to Richard Spruce

[Letters to R. Spruce] Correspondence 1842-1890. *At Royal Botanic Gardens, Kew, Library & Archives, with the Spruce Papers.*
  Includes letters from: Sir W.J. Hooker, G. Bentham, W. Mitten, A. Destruge, J. Miers and E. Bescherelle, with some draft replies.

Other papers

Drawings of scenery and natives in Amazonian Brazil, c. 1850 by R. Spruce. *At the Royal Society of London*, ms. 236.

Maps of the Amazon region. ms. 11 ms. maps and 26 printed maps. 1850s. *At the Manchester Museum.*

Notes for introduction to "Hepaticae amazonicae et andinae." 1884? *At the Royal Botanic Gardens, Kew, Library & Archives.* Bound with R. Spruce letters to W. Borrer, leaves 84-107.

Notes on the possible acclimatization of Europeans in tropical South America. [after 1865]. *At the Linnean Society of London, with Spruce's list of botanical excursions.*
  There is a ms. rough draft of this at RBG Kew, in R. Spruce's letters to W. Borrer, doc. 112-113.

[Personal papers. 1817-1893]. *At Manchester Central Library (Archives Dept.)*, MSF 925.8 SP1.
  Includes: copy of register of Spruce's baptism, certificates from learned societies, records of expenses...1860-61, letters from the India Office concerning his pension, and from the Linnean Society of London on his election to Associateship, 1893, and a draft reply; sketches; lists of illustrations.

Related material

Royal Geographical Society. [Correspondence files of incoming letters (few for 1863-1869), including A.R. Wallace 1853-1908, and Edward Whymper, 1875-1908]. *At the Royal Geographical Society, London.*

Wallace, A.R. Sketches of the palms of the Amazon, with an account of their uses and distribution, published 1853; 21 drawings of Amazon palms, etc. *At the Linnean Society of London.*

Wallace, A.R. Letters to Matthew B. Slater, 1886-1909, concerning Wallace's editing of Spruce's *Notes of a botanist* and related matters, with some draft replies and transcripts. *At Manchester Central Library (Archives Dept.)* MSF 925 WA1.

[G. Stabler correspondence on Richard Spruce, with I.B. Balfour and P.N. Fraser]. 2 letters. *At the Royal Botanic Garden, Edinburgh.*

27

Bibliography
of Richard Spruce

M. R. D. Seaward

Bibliography of Richard Spruce

M.R.D. Seaward

University of Bradford, UK

This bibliography of works by Richard Spruce, and works about him, attempts to be comprehensive in terms of Spruce's published output; it rectifies the numerous errors and omissions in Stabler's (1894) obituary of Spruce and in Wallace's (1908) introduction to *Notes of a Botanist on the Amazon and Andes,* which have unfortunately been perpetuated in subsequent works. It is accepted that further ms. works which Spruce intended for publication may come to light, and that further biographical accounts in works by other authors could no doubt be added to the list below.

In considering publications *about* Spruce and his work, it would clearly be difficult to list all publications discussing the plant specimens he collected; however, papers by George Bentham, M.J. Berkeley and others on their initial studies of Spruce specimens have been included, as part of the historical *fonds* from which subsequent work derives.

Titles have been cited exactly as they appear in print; however, where occasionally necessary, scientific nomenclature has been italicised, and initial capitalisation of specific epithets and punctuation before authorities silently amended to accord with modern practice.

The author is greatly indebted to Miss Gina Douglas (Librarian of the Linnean Society), Ms Sylvia FitzGerald (Royal Botanic Gardens, Kew), Dr Roy Watling (Royal Botanic Garden, Edinburgh), Dr John Dickenson (University of Liverpool), Dr Marshall Crosby (Missouri Botanical Garden), Dr Nigel Smith (University of Florida), and Prof. R.E. Schultes and Dr Gustavo Romero (Harvard University) for valuable bibliographical information.

Works by Spruce

1841   Three days on the Yorkshire Moors. *Phytologist* **1**: 101-104.

1842   Discovery of *Leskea pulvinata* Wahl. *Phytologist* **1**: 189.

1842   List of mosses etc. collected in Wharfedale, Yorkshire. *Phytologist* **1**: 197.

1842   Note on *Didymodon flexicaulis. Phytologist* **1**: 197-198.

1842   Mosses near Castle Howard. *Phytologist* **1**: 198.

1842   *Bryum pyriforme. Phytologist* **1**: 429.

1842   On the folia accessoria of *Hypnum filicinum* Lin. *Phytologist* **1**: 459-461.

1843   A list of mosses and Hepaticae collected in Eskdale, Yorkshire. *Phytologist* **1**: 540-544.

1844   Note on *Carex paradoxa. Phytologist* **1**: 842.

1844   Note on *Carex axillaris. Phytologist* **1**: 842-843.

1844   Note on *Veronica triphyllos. Phytologist* **1**: 843.

1844   *Veronica buxbaumii. Phytologist* **1**: 843.

1845   On the branch-bearing leaves of *Jungermannia juniperina* (Sw.). *Phytologist* **2**: 85-86.

1845   A list of the Musci and Hepaticae of Yorkshire. *Phytologist* **2**: 147-157.

1845   On several mosses new to the British flora. *Lond. J. Bot.* **4**: 169-195.

1846   The Musci and Hepaticae of Teesdale. *Trans. Bot. Soc. Edinb.* **2**: 65-89.

1846   Notes on the botany of the Pyrenees, in a letter addressed to the editor. *Lond. J. Bot.* **5**: 134-142, 345-350, 417-429, 535-548.

1847   *Hepaticae Pyrenaicae, quas in Pyrenaeis centralibus occidentalibusque, nec non in Agro Syrtico, A.D. 1845-6.* Numbers 1-77. Londini. [Exsiccatae].

1847   *Musci Pyrenaici, quos in Pyrenaeis centralibusque occidentalibusque, nec non in Agro Syrtico, A.D. 1845-6 decerpsit Rich. Spruce.* Fascis I, numbers 1-160; Fascis II, numbers 161-331. Londini. [Exsiccatae].

1849   Mr Spruce's voyage to Pará. *Hooker's J. Bot.* **1**: 344-347.

1850   The Musci and Hepaticae of the Pyrenees. *Trans. Bot. Soc. Edinb.* **3**: 103-216, t.3. [Read 11th January 1849].

         [Also published in parts in advance of the above in *Ann. Mag. Nat. Hist.,* ser 2, **3**: 81-106, 269-293, 358-380, 478-503, t.3; **4**: 104-120 (1849)]

[c.1850]   *Lichenes Pyrenaei.* Collegit R. Spruce: determinavit Churchill Babington.

         [Labels to undetermined number of specimens in BM and elsewhere; cited in Lindau, G. & Sydow, P. (1909) *Thesaurus litteraturae mycologicae et lichenologicae.* Lipsiis: Borntraeger, Vol. 2. p. 564]

1850   Botanical excursion on the Amazon. *Hooker's J. Bot.* **2**: 65-70.

         [List of vegetable curiosities sent by Spruce to the Kew Museum, with additional information provided by Spruce. *Hooker's J. Bot.* **2**: 70-76]

1850   Voyage up the Amazon river. *Hooker's J. Bot.* **2**: 173-178.

1850   Journal of an excursion from Santarém, on the Amazon river, to Obidos and the Rio Trombetas. *Hooker's J. Bot.* **2**: 193-208, 225-232, 266-276, 298-302.

1851   Extracts of letters from Richard Spruce, Esq., written during a botanical mission on the Amazon. *Hooker's J. Bot.* **3**: 84-89, 139-146.

1851   Copy of a letter addressed by Mr. Spruce to G. Bentham, Esq., dated Santarém, Rio das Amazonas, Sept. 10, 1850. *Hooker's J. Bot.* **3**: 239-248.

1851   Journal of a voyage from Santarém to the Barra do Rio Negro. *Hooker's J. Bot.* **3**: 270-278, 335-343.

1852   Intelligence of Mr. Spruce, in a letter to G. Bentham, Esq. *Hooker's J. Bot.* **4**: 278-281.

1852   Copy of a letter from Mr. Spruce, addressed to Mr. John Smith, Royal Gardens, Kew, dated Falls of S. Gabriel, Rio Negro, Dec. 28, 1851. *Hooker's J. Bot.* **4**: 282-285.

1852   Letter from Mr. Spruce to George Bentham, Esq. *Hooker's J. Bot.* **4**: 305-312.

1853   Edible fruits of the Rio Negro, South America. *Hooker's J. Bot.* **5**: 183-187.

1853   Botanical objects communicated to the Kew Museum, from the Amazon River, in 1851. *Hooker's J. Bot.* **5**: 169-177.

1853   Botanical objects communicated to the Kew Museum, from the Amazon River, in 1851 and 1852. *Hooker's J. Bot.* **5**: 238-247.

1853-54   Journal of a voyage up the Amazon and Rio Negro. *Hooker's J. Bot.* **5**: 187-192, 207-215; **6**: 33-42, 107-111.

1854   Extract of a letter relating to vegetable oils, etc. *Hooker's J. Bot.* **6**: 333-337.

1855   Journal of a botanical voyage up the Amazon, Rio Negro, and to the Casiquiare. *Hooker's J. Bot.* **7**: 1-8.

1855   Note on the India-rubber of the Amazon. *Hooker's J. Bot.* **7**: 193-196.

1855   Botanical objects communicated to the Kew Museum, from the Amazon or its tributaries, in 1853. *Hooker's J. Bot.* **7**: 209-210, 245-252, 273-278.

1855   Sarsaparilla. Extract from a letter from Mr Spruce, dated Rio Negro, February 5, 1855. *Hooker's J. Bot.* **7**: 214-215.

1855   Note sur le caoutchouc de la rivière des Amazones. *J. Pharm. Chimie*, ser. 3, **28**: 382-284.

1855   Mr Spruce's voyage up the Amazon and its tributaries [letter dated March 11, 1855]. *Hooker's J. Bot.* **7**: 281-282.

1855   Note on Clusiaceae. *Hooker's J. Bot.* **7**: 347-348.

1856   Note on the India-rubber of the Amazon. *Pharm J.* **15**: 117-119. [Reprint of letter by Spruce from Barra do Rio Negro dated 9 February 1855 first published in *Hooker's J. Bot.*]

1856   Mr Spruce in Peru. *Hooker's J. Bot.* **8**: 177-181.

1859   On five new plants from eastern Peru. *J. Proc. Linn. Soc., Bot.* **3**: 191-204.

     [Namely *Wettinia illaqueans*, a new palm from the Peruvian Andes; *Discanthus*, a new genus of Cyclanthaceae; *Yangua tinctoria*, a new genus of Bignoniaceae; *Capirona*, a new genus of Rubiaceae; *Erythrina amasisa*, a new species with follicular pods].

1860   [recte 1859] On *Leopoldinia piassaba* Wallace. *J. Proc. Linn. Soc., Bot.* **4**: 58-63.

1860   Los cerros de Llanganati ... memoria presentada a la Sociedad Geográfica de Londres ... 1860. [English version publ. 1861. Spanish ms. in RBG Kew Archives, in Spruce papers – Letters to W. Borrer leaf 108]

1860   Notes of a visit to the Cinchona forests on the western slope of the Quitenian Andes. *J. Proc. Linn. Soc., Bot.* **4**: 176-192.

[1860]   *Report on expedition to procure seeds and plants of* Cinchona succirubra, *or Red bark tree.* London: HMSO (House of Commons Paper 865). 6pp. [Reprinted in *East India (Chinchona Plant)*, "Blue Book I", 1852-1863: 60-64].

1861   [recte 1860] On the mode of branching of some Amazon trees. *J. Proc. Linn. Soc., Bot.* **5**: 3-14.

1861   [recte 1860] Mosses of the Amazon and Andes. *J. Proc. Linn. Soc., Bot.* **5**: 45-51.

1861   *Report on the expedition to procure seeds and plants of the* Cinchona succirubra *Pavon, or Red-bark tree.* India Office. Pp. 112, map. [With a note on the map by C.R. Markham; see also Spruce (1908) *Notes of a Botanist on the Amazon and Andes* **2**: 261-293]

1861   [recte 1862] *Report on the expedition to procure seeds and plants of the* Cinchona succirubra *or Red Bark tree.* London: Eyre & Spottiswoode, for HMSO, Pp. 112, map.

1861   On the mountains of Llanganati, in the eastern Cordillera of the Quitonian Andes, illustrated by a map constructed by the late Don Atanasio Guzman. *Jl R. Geogr. Soc.* **31**: 163-184.

1862   *On the mountains of Llanganati, in the eastern Cordillera of the Quitonian Andes, illustrated by a map constructed by the late Don Atanasio Guzman.* London: W. Clowes. Pp. 22.

[1862]   *Note on the cultivation of Chinchonae.* London: HMSO (House of Commons Paper 2954). 2 pp. [Reprinted in *East India (Chinchona Plant)*, "Blue Book I", 1852-1863: 227-228]

[1862]   Note by Richard Spruce on the tradition respecting Ursua and Aguirre, amongst the Indians of the river Huallaga. 3 pp.

1863   From R. Spruce, Esq., to the Under Secretary of State for India. *East India (Chinchona Plant)*, "Blue Book I", 1852-1863: 58-59.

1863   From R. Spruce, Esq., to Clements Markham, Esq. *East India (Chinchona Plant)*, "Blue Book I", 1852-1863: 59.

1863   Mr Spruce's report on the expedition to procure seeds and plants of the *Cinchona succirubra*, or red bark tree, to the Under Secretary of State for India, 3rd January 1862. *East India (Chinchona Plant)*, "Blue Book I", 1852-1863: 65-118, 2 maps.

1864   *Notes on the Valleys of Piura and Chira, in northern Peru, and on the cultivation of cotton therein.* London: Eyre & Spottiswoode, for HMSO. Pp. 81. [See also Spruce (1908). *Notes of a Botanist on the Amazon and Andes* **2**: 327-341]

1864   On the River Purus, a tributary of the Amazon. In: *The Travels of Pedro de Cieza de Leon, A.D. 1532-50, contained in the first part of his Chronicle of Peru* (C.R. Markham, ed.): 339-351. London: Hakluyt Society. [As a note to Chapter 95; reprinted as a pamphlet with renumbered pagination]

1865   On the physical geography of the Peruvian coast valleys of Chira and Piura, and the adjacent deserts. *Rep. Br. Ass. Advmt Sci.* **1864**: 148.

1865   On the River Purus. *Rep. Br. Ass. Advmt Sci.* **1864**: 148.

1865   Note on the volcanic tufa of Latacunga, at the foot of Cotopaxi; and on the Cangaua, or volcanic mud, of the Quitenian Andes. *Lond. Edinb. Dubl. Phil. Mag.* **29**: 401.

1865   Note on the volcanic tufa of Latacunga, at the foot of Cotopaxi; and on the Cangaua, or volcanic mud, of the Quitenian Andes. *Q. Jl Geol. Soc. Lond.* **21**: 243-250.

1865   Beal-fires. *The Reader* **6**: 569.

1866 *The White Island. An apologue on sabbatarianism, in the style of Swift.* The English Leader, no. 63. London.

1867 *Catalogus muscorum fere omnium quos in Terris Amazonicis et Andinis, per annos 1849-1860, legit Ricardus Spruceus.* Londini: printed by E. Newman. Pp. 22.

  [Catalogue of exsiccata, *Musci Amazonici et Andini. Legit. Ric. Spruce, det. W. Mitten*, nos. 1-1518, distributed in 1866]

1868 Notes on some insect and other migrations observed in Equatorial America. *J. Linn. Soc., Zool.* **9**: 346-367. [See also Spruce (1908) *Notes of a Botanist on the Amazon and Andes* **2**: 353-383]

1869 [with Joaquim Correa de Mello] Notes on Papayaceae. *J. Linn. Soc., Bot.* **10**: 1-15, t.1. [Written by Spruce, with added or amended observations by de Mello provided in brackets]

1869 *Palmae Amazonicae, sive enumeratio palmarum in itinero suo per regiones Americae aesquatoriales lectarum. J. Linn. Soc., Bot.* **11**: 65-183. [See also: Regelmässiger Wechsel in der Entwickelung diclinischer Blüthen. *Bot. Ztg* **27**: 664-666 (1869)]

1871 [recte 1870] On the fertilisation of grasses. *Am. Nat.* **4**: 239-241.

1874 Personal experiences of venomous reptiles and insects in South America. *Ocean Highways: Geogrl Rev.*, n.s. **1**: 135-146.

1874 On some remarkable narcotics of the Amazon Valley and Orinoco. *Ocean Highways: Geogrl Rev.*, n.s. **1**: 184-193. [See also Spruce (1908) *Notes of a Botanist on the Amazon and Andes* **2**: 413-455]

[c. 1874] *Lichenes Amazonici et Andini.* [Exsiccata, distribution of which commenced in or before 1874 (see Sayre 1975: 401-402); 851 non-consecutive numbers (see Lynge 1915-22: 494-501). Some unnumbered lichens also distributed as *Lichenes Amazonici, coll. R. Spruce, 1849, determinn. C. Montagne et C. Babington*]

1876 On *Anomoclada*, a new genus of Hepaticae, and on its allied genera, *Odontoschisma* and *Adelanthus. J. Bot. Lond.* **4**: 129-126, 161-170, 193-203, 230-235, t.2.

1879 *Linnaea borealis* in Yorkshire. *J. Bot., Lond.* **17**: 184.

1879 *Hypnum (Brachythecium) salebrosum* Hoffm., as a British moss. *J. Bot., Lond.* **17**: 305-307.

1880-81 Musci praeteriti: sive de Muscis nonnullis adhuc, praetervisis vel confusis, nunc recognitis. *J. Bot. Lond.* **18**: 289-295, 353-361; **19**: 11-18, 33-40.

1881 On *Marsupiella stableri* (n.s.) and some allied species of European Hepaticae. *Revue Bryol.* **8**: 89-104.

1881 The morphology of the leaf of *Fissidens. J. Bot., Lond.* **19**: 98-99.

1882 *On* Cephalozia *(a genus of Hepaticae). Its subgenera and some allied genera.* Malton: printed by J.W. Slater for the author. Pp. vi + 96 + [3]. [See review in *J. Bot., Lond.* **21**: 183-187 (1883)]

1882 Liverworts (Hepaticae) of the East Riding. *Trans. Yorks. Nat. Un.* **4**: 62-63.

1884-85   Hepaticae Amazonicae et Andinae quas in itinere suo per tractus montium et fluviorum Americae aequinoctialis ... &c. *Trans. Bot. Soc. Edinb.* **15**: i-xii, 1-588 [-590], t. 22. [Published in two parts: first in April 1884 as **15**: i-xii, 1-308; the second in November 1885 as **15** (2): 309-588[-590], t.22]

1885   *Hepaticae of the Amazon and of the Andes of Peru and Ecuador.* Pp. xii + 590, t.22. London: Trubner.

[As above, but with half title page of prefatory note which appeared on inside cover of *Trans. Bot. Soc. Edinb.* **15** (2). Reprinted in 1984 with variant title *Hepaticae of the Amazon and the Andes of Peru and Ecuador* and an introduction and index with updated nomenclature by Barbara M. Thiers, as *Contr. N.Y. Bot. Gdn.* **15**: I-XVI, i-xii, 1-590, t.22, (1)-(14)]

1886   Précis d'un voyage d'exploration botanique dans l'Amérique équatoriale, pour servir d'introduction provisoire à son ouvrage sur les Hépatiques et l'Amazon et des Andes. *Revue Bryol.* **13**: 61-79.

[Reprint cover bears different title: 'Voyage de R. Spruce dans l'Amérique ÉEquatoriale, pendant les années 1849-1864', and is paginated pp. 1-20]

1887   *Lejeunes holtii*, a new hepatic from Killarney. *J. Bot., Lond.* **25**: 33-39, 72-82, t.1.

1887   On a new Irish hepatic (*Radula holtii*). *J. Bot., Lond.* **25**: 209-211.

1888   Hepaticae in Província Rio Janeiro, a Gloziou [=Glaziou] lectae, a R. Spruce determinatae. *Revue Bryol.* **15**: 33-34.

1888   Hepaticae Paraguayenses, Balansa lectae, R. Spruce determinatae. *Revue Bryol.* **15**: 34-35.

1889   *Lejeunea rossettiana* Massal. *J. Bot., Lond.* **27**: 337-338.

1889   Hepaticae novae Americanae tropicae et aliae. *Bull. Soc. Bot. Fr.* **36** suppl. Congrès de Botanique, Paris 1889: clxxxix-ccvi.

1889   [with E. Bescherelle] Hépatiques nouvelles de Colonies françaises. *Bull. Soc. Bot. Fr.* **36** suppl. Congrès de Botanique, Paris 1889: clxxvii-clxxix, t.5.

1890   Hepaticae Bolivianae, in Andibus Boliviae orientalis annis, 1885-6, a cl. H.H. Rusby lectae. *Mem. Torrey bot. Club* **1**: 113-140. [See review by W.H.P[earson] in *J. Bot., Lond.* **28**: 252-253 (1890)]

1892   *Hepaticae Spruceanae, Amazonicae et Andinae, annis 1849-1860 lectae.* Malton: printed for the author. [See review *Bot. Gaz.* **18**: 112-113 (1893)]

1895   Hepaticae Elliottianae, insularis Antillanis Sti Vincentii et Dominica a clar. W.R. Elliott, annis 1891-92, lectae, Ricardo Spruce determinatae. *J. Linn. Soc., Bot.* **30**: 331-372, t.11.

Publications about Spruce and his work

Angel, R. (1978) Richard Spruce, botanist and traveller, 1817-1893. An exhibition in The Orangery, Royal Botanic Gardens, Kew. *Hortulus Aliquando* **3**: 49-53.

Anon. (1849) Mr Spruce's intended voyage to the Amazon River. *Hooker's J. Bot.* **1**: 20-21.

Anon. (1850) Mr Spruce's journey. *Hooker's J. Bot.* **2**: 158.

Anon. (1854) Mr Spruce's South American plants. *Hooker's J. Bot.* **6**: 94.

Anon. (1855) Mr Spruce's ascent of the Amazon to Peru. *Hooker's J. Bot.* **7**: 380.

Anon. (1856) Mr Spruce's collections [from the vicinity of Tarapotu, in Peru]. *Hooker's J. Bot.* **8**: 379.

Anon. (1857) Mr Spruce at Tarapota. *Hooker's J. Bot.* **9**: 310-311.

Anon. (1863) *Statement of the results of Mr Richard Spruce's travels in the Valley of the Amazon, and in the Andes of Peru and Ecuador.* [With: Note by Mr. Bentham, on Mr. Spruce's services to botany.] London. Pp. 7.

[Privately printed to support application for a pension for Spruce; see also letter from C.R. Markham and reprinted statement ... &c. in *East India (Chinchona Plant)*, "Blue Book II", 1863-1866: 247-249 & 250-251]

Anon. (1864) Botanical explorations of Mr Richard Spruce. *J. Bot., Lond.* **2**: 199-201. [Most likely written by the editor, B. Seemann]

Anon. (1883) Notices of books. On *Cephalozia* ... &c. *Naturalist* **8**: 156-158.

Anon. (1895) [Richard Spruce: obituary]. *Proc. Linn. Soc.* **1893-1894**: 35-37.

Balfour, I.B. (1900) Richard Spruce. *Ann. Bot.* **14**: xi-xiv, t.1.

Bentham, G. (1850) Report on the dried plants collected by Mr. Spruce in the neighbourhood of Pará in the months of July, August and September 1849. *Hooker's J. Bot.* **2**: 209-212; 233-244.

Bentham, G. (1851) Second report on Mr. Spruce's collections of dried plants from North Brazil. *Hooker's J. Bot.* **3**: 111-120; 161-166; 191-200; 366-373.

Bentham, G. (1852) Second report on Mr. Spruce's collections of dried plants from North Brazil [continued]. *Hooker's J. Bot.* **4**: 8-18.

Bentham, G. (1853) On some genera and species of Brazilian Rubiaceae. *Hooker's J. Bot.* **4**: 229-236. [Includes description of new genus *Sprucea*, p. 229-230]

Bentham, G. (1854) Notes on North Brazilian Gentianeae, from the collections of Mr. Spruce and Sir Robert Schomburgk. *Hooker's J. Bot.* **6**: 193-204.

Bentham, G. (1854) On the North Brazilian Euphorbiaceae in the collection of Mr. Spruce. *Hooker's J. Bot.* **6**: 321-333, 363-377.

Bentham, G. (1854) On *Henriquezia verticillata*, Spruce: a new genus of Bignoniaceae, from the Rio Negro, in North Brazil. *Hooker's J. Bot.* **6**: 337-339.

Bentham, G. (1855) On the South American Triurideae and leafless Burmanniaceae from the collections of Mr. Spruce. *Hooker's J. Bot.* **7**: 8-17.

[Bentham, G.] (1855) Mr Spruce's plants of the Amazon River and its tributaries. *Hooker's J. Bot.* **7**: 31.

Berkeley, M.J. (1856) Decades of Fungi. Decades LI-LIV, [-LXII]. Rio Negro fungi [collected by R. Spruce]. *Hooker's J. Bot.* **8**: 129-144; 169-177; 193-200; 233-241; 272-280.

B[oulger], G.S. (1882-1900) Richard Spruce. In: *Dictionary of National Biography* **53**: 431-432.

Clokie, H.N. (1964) *An account of the herbaria of the Department of Botany in the University of Oxford.* Oxford: Oxford University Press. [p. 247]

Cutright, P.R. (1940) *The great naturalists explore South America.* New York: Macmillan. Pp. 340.

Desmond, R. (1977) *Dictionary of British and Irish botanists and horticulturists.* London: Taylor & Francis. [p. 578] and Rev. ed., 1994, pp. 647-648.

Ernst, A. (1867) El Doctor Ricardo Spruce. *El Federalista Caracas* **4** (9): [2-3].

Furneaux, R. (1969) *The Amazon: the story of a great river.* London: Hamish Hamilton.

Gepp, A. (1894) In memory of Richard Spruce. *J. Bot., Lond.* **32**: 50-53.

Hawksworth, D.L. & Seaward, M.R.D. (1977) *Lichenology in the British Isles 1568-1975.* Richmond: Richmond Publishing. [p. 154]

Hemming, J. (1987) *Amazon frontier: the defeat of the Brazilian Indians.* London: Macmillan; Cambridge, Mass.: Harvard University Press.

[11 entries for Spruce in index, plus biographical note on pp. 499-500.]

Huber, O. & Wurdack, J.J. (1984) History of botanical exploration in Território Federal Amazonas, Venezuela. *Smithson. Contr. Bot.* **56**: i-iii, 1-83.

H[usnot], T. (1894) Nécrologie [Richard Spruce]. *Revue bryol.* **21**: 46-47.

King, G. (1876) *A Manual of Cinchona cultivation in India.* Calcutta: Office of the Superintendant of Government Printing. 80 pp.

Leighton, W.A. (1866) Lichenes Amazonici et Andini lecte a Domino Spruce. *Trans. Linn. Soc. Lond.* **25**: 433-460, t.1. [See also Lynge (1915-22)]

Lindley, J. (1852-1859) Folia orchidaceae. London. [includes 114 descriptions of new species collected by Spruce & 1 new variety; for details, see Romero in this volume, p. 206].

Lynge, B. (1915-22) Index specierum et varietatum lichenum quae collectionibus 'Lichenes exsiccati' distributae sunt. Kristiania: A.W. Broggers. [pp. 495-601] [Issued with separate pagination from *Nyt. Mag. Naturvid.* **53-60**: Pars I (1915-19), 559 pp.; Pars II (1920-22), 316 pp. See also Leighton (1866)]

McGill, H.M. (1960-61) The case of the missing journal. *Manchester Review* **9**: 124-128.

MacKinder, B.A., Owen, P.E. & Simpson, K. (1990) *Richard Spruce's legumes from the Amazon.* Kew: Royal Botanic Gardens. Pp. iv + 31.

Markham, C.R. (1880) *Peruvian bark: a popular account of the introduction of chinchona cultivation into British India.* London: John Murray. Pp. xxiv + 550. [Many references to Spruce, mainly chapter 20 entitled 'Dr. Spruce's expedition to procure plants and seeds of the "red bark", or *C. succirubra*' pp. 217-227]

Markham, C.R. (1894) Richard Spruce. *Geogr. J.* **3**: 245-247.

Mitten, W. (1869) Musci Austro-Americani. *J. Linn. Soc., Bot.* **12**: 1-659. [Reprinted in 1982 as Monographs in Systematic Botany from the Missouri Botanical Garden, vol. 7, with additional title-page and including 1907 obituary of Mitten by E.M. Holmes from *Proc. Linn. Soc.* **119**: 49-54]

Muller, J. (1892) Lichenes epiphylii spruceani. *J. Linn. Soc. Bot.* **29**: 322-333.

Nylander, W. (1874) Animadversiones circa Spruce *Lichenes Amazonicos et Andinos. Flora* (*Regensburg*) N.R. **32**: 70-73.

Pearson, M.B. (1990) Richard Spruce's "list of botanical excursions". *Linnean* **6** (2): 18-20.

Prance, G.T. (1971) An index of plant collectors in Brazilian Amazonia. *Acta Amazonica* **1** (1): 25-65.

Reichenbach, H.G.f. (1873) Zum geographischen Verstandniss der amerikanischen Reisepflanzen des Herrn Dr. Spruce. *Bot. Ztg.* **31**: 27-28.

Renner, S.S. (1993) A history of botanical exploration in Amazonian Ecuador, 1739-1988. *Smithson. Contr. Bot.* **82**: i-iii, 1-39.

Richards, P.W. (1994) Richard Spruce, the man. *Bull. Brit. Bryol. Soc.* **63**: 54-58.

Sandeman, C. (1949) Richard Spruce: portrait of a great Englishman. *Jl R. Hort. Soc.* **74**: 531-544.

Sayre, G. (1971) Cryptogamae exsiccatae – an annotated bibliography of exsiccatae of algae, lichenes, hepaticae, and musci. IV. Bryophyta. *Mem. N.Y. Bot. Gdn.* **19**: 175-276. [Richard Spruce p. 257-258].

Sayre, G. (1975) Cryptogamae exsiccatae – an annotated bibliography of exsiccatae of algae, lichens, hepaticae, and musci. V. Unpublished exsiccatae. I. Collectors. *Mem. N.Y. Bot. Gdn.* **19**: 277-423. [Richard Spruce p. 401-402].

Schultes, R.E. (1951) Plantae Austro-americanae VII. *Bot. Mus. Leafl. Harv. Univ.* **15**(2): 29-78.

Schultes, R.E. (1953) Richard Spruce still lives. *Northern Gardener* **7**: 20-27, 55-61, 87-93, 121-125. [Also issued as repaginated reprint, pp. 1-27]

Schultes, R.E. (1968) Some impacts of Spruce's Amazon exploration on modern phytochemical research. *Rhodora* **70**: 313-339. [Reprinted, with minor changes, from *Ciencia e Cultura* **20**: 37-49 (1968)]

Schultes, R.E. (1970) The history of taxonomic studies in *Hevea*. *Regnum Vegetabile* **71**: 229-293.

Schultes, R.E. (1978) An unpublished letter by Richard Spruce on the theory of evolution. *Biol. J. Linn. Soc.* **10**: 159-161.

Schultes, R.E. (1978) Richard Spruce and the potential for European settlement of the Amazon: an unpublished letter. *Bot. J. Linn. Soc.* **77**: 131-139.

Schultes, R.E. (1978) Richard Spruce still lives. *Hortulus Aliquando* **3**: 13-47.

Schultes, R.E. (1983) Richard Spruce: an early ethno-botanist and explorer of the northwest Amazon and northern Amazon and northern Andes. *J. Ethnobiol.* **3**: 139-147.

Schultes, R.E. (1985) Several unpublished ethnobotanical notes of Richard Spruce. *Rhodora* **87**: 439-441.

Schultes, R.E. (1987) Still another unpublished letter from Richard Spruce on evolution. *Rhodora* **89**: 101-106.

Schultes, R.E. (1990) Notes on difficulties experienced by Spruce in his collecting. *Rhodora* **92**: 42-44.

Schultes, R.E. (1990) Margaret Mee and Richard Spruce. *Naturalist* **115**: 146-148.

Schultes, R.E., Holmstedt, B. & Lindgren, J.-E. (1969) De plantis toxicariis e mundo novo tropicale commentationes III. Phytochemical examination of Spruce's original collection of *Banisteriopsis caapi*. *Bot. Mus. Leafl. Harv. Univ.* **22** (4): 121-164.

Schultes, R.E. & Raffauf, R.F. (1992) A rare report of an intoxicating snuff from the Amazon. *Kew Bull.* **47**: 743-744.

Schuster, R.M. (1982) Richard Spruce (1817-1893): a biographical sketch and appreciation. *Nova Hedwigia* **36**: 199-208.

Scott, L.I. (1961) Bryology and bryologists in Yorkshire. *Naturalist* **86**: 155-160.

Seaward, M.R.D. (1980) Two letters of bryological interest from Richard Spruce to David Moore. *Naturalist* **105**: 29-33.

Seaward, M.R.D. (1995) Spruce's diary. *Linnean* **11**: 17-19.

Seaward, M.R.D. (1997) Richard Schultes and the botanist-explorer Richard Spruce (1817-1893). In: [Festschrift for Richard Schultes]. Portland, OR; Dioscorides Press. (In press)

Sheppard, T. (1909) A Yorkshire botanist: Richard Spruce (1817-1893). *Naturalist* **34**: 45-48.

Slater, M.R. (1906) The mosses and hepaticae of North Yorkshire. In: *North Yorkshire: studies of its botany, geology, climate, and physical geography* (J.G. Baker). 2nd edition. *Trans. Yorks. Nat. Un.*, bot. ser. **3**: i-xvi, 417-671.

Sledge, W.A. (1971) Richard Spruce. *Naturalist* **96**: 129-131.

Sledge, W.A. & Schultes, R.E. (1988) Richard Spruce: a multi-talented botanist. *J. Ethnobiol.* **8**: 7-12.

Smith, A. (1990) *Explorers of the Amazon.* London: Viking. [Particularly Chapter 8, 'Spruce & Wickham – explorers extraordinary', pp. 251-284]

Spruce, R. (1908) *Notes of a botanist on the Amazon and Andes* (ed. A.R. Wallace). 2 volumes. London: Macmillan, Pp. lii + 518, t.3; xii + 542, t.4.

[Reprinted edition (1970) with a new foreword by R.E. Schultes. 2 volumes. New York: Johnson Reprint Corp. Pp. x + lii + 518, t.3; xii + 542, t.4]

Spruce, R. (1938) *Notas de un Botânico sobre el Amazonas y los Andes* (ed. A.R. Wallace). Quito: Publicaciones de la Universidad Central. Pp. 422. [Translation by G. Salgado of Volume I of the 1908 edition, without figures & maps]

Stabler, G. (1894) Obituary notice of Richard Spruce, Ph.D. *Trans. Bot. Soc. Edinb.* **20**: 99-109.

Stafleu, F.A. & Cowan, R.S. (1985) *Taxonomic literature. A selective guide to botanical publications and collections with dates, commentaries and types.* 2nd ed. Volume 5. Utrecht: Bohn, Scheltema & Holkema. [pp. 816-820]

Stephani, F. (1894) Richard Spruce. *Botanisches Centralblatt* **57**: 370-374.

Sterling, T. (1972) *The Amazon.* Amsterdam: Time-Life Books. [pp. 124-125]

Thiers, B.M. (1992) Indices to the species of mosses and lichens described by William Mitten. *Mem. N.Y. Bot. Gdn* **68**: i-iv, 1-113.

Underwood, L.M. (1893) A notable collection of Hepaticae. *Bot. Gaz.* **18**: 112-113.

Urban, I. (1906). Spruce, Richard (1817-1893). In: Martius, C.F.P.de. Vitae itineraque collectorum botanicorum, notae collaboratorum biographicae &c. *Flora Brasiliensis*, **1** (1): 113-116, Monachii: Oldenbourg. [reprinted by J. Cramer, Weinheim, 1965]

Von Hagen, V.W. (1949) *South America called them.* London: Robert Hale. Pp. xiv + 401, t.28. [Part 4, pp. 291-376, 368-387, devoted to Spruce; an article adapted from chapter, entitled 'The great mother forest: a record of Richard Spruce's days along the Amazon' appeared in *J. N.Y. Bot. Gdn* **45**: 73-80 (1944)]

W[allace], A.R. (1894) Richard Spruce, Ph.D., F.R.G.S. *Nature* **49**: 317-319.

Whiffen, T. (1915) *The Northwest-Amazons; notes of some months spent among cannibal tribes.* London: Constable.

Wilkinson, H.J. (1907) Historical account of the herbarium of the Yorkshire Philosophical Society and the contributors thereto. Richard Spruce. *Rep. Yorks. Phil. Soc.* **1907**: 59-67.

Appendix

Richard Spruce (1817–1893), botanist and explorer: a commemorative conference

The Linnean Society of London
Annual Regional Meeting
York, 20-22 September 1993

Richard Spruce (1817–1893), botanist and explorer: a commemorative conference

The Linnean Society of London
Annual Regional Meeting,
York, 20–22 September 1993

Sponsorship from the Linnean Society of London, the National Grid Company, the Leeds Philosophical & Literary Society, CIBA-Geigy and a number of other institutions, particularly in North America, is gratefully acknowledged.

The Linnean Society's Commemorative Conference to celebrate the life and extraordinary achievements of the Yorkshire botanist-explorer Richard Spruce was held at York on the hundredth anniversary of his death. More than sixty participants (see below), including many from North America and representatives from each of the South American countries visited by Spruce, enjoyed a varied programme of lectures, exhibitions and visits centred on the University of York, 20-22 September 1993. The President of the Linnean Society, Professor J.G. Hawkes, hosted the Conference, which included 26 stimulating lectures and posters, and a magnificent ethnobotanical exhibition of Spruce materials prepared by the Royal Botanic Gardens, Kew: almost all these contributions and a few additional papers are published in this commemorative volume.

On the first evening, after a full day of lecture and poster presentations, those attending the Conference were treated to a memorable public lecture at the Yorkshire Museum given by Professor Ghillean Prance, who described himself as 'a contemporary botanist in the footsteps of Richard Spruce'; he skilfully illustrated quotations from Spruce's journals with his own photographic slides to show how it was still possible to experience many of the things so vividly described by Spruce. Afterwards, the participants and guests enjoyed a celebratory buffet at the Museum, presided over by the Sheriff of York and his lady.

The following day, Conference participants were transported by coach to the Castle Howard estate, stopping first at Coneysthorpe where Spruce spent most of his life after returning to England. After a short introduction to the estate, given by The Hon. Simon Howard on the village green, Spruce's cottage was inspected internally by a few participants by kind permission of its present occupant. The cottage is outwardly little changed over the past century other than by the addition of a commemorative plaque erected in 1971.

The party moved on to Terrington, where coffee was kindly provided by the staff of the Terrington Hall Preparatory School, and then attended a Remembrance Service in the adjacent church conducted by the Rector, the Rev. Edwin Chapman; prayers at Spruce's graveside were followed by a service inside the church which included addresses by Professor Richard Schultes and Mr Winston Spruce, who also provided a tape-recording of Spruce's hymn-tune 'Raywood'.

After a tour of Castle Howard by kind permission of its owner, which included an opportunity to view a very rare and ornate South American hammock posted to a Howard forbear by Spruce, Conference participants and guests sat down to a magnificent luncheon in the Grecian Room presided over by The Hon. Simon Howard and his wife, who paid participants the compliment of gracing each table with a piece of Georgian silverware. In the afternoon, papers were given in a lecture room at Castle Howard, the setting further enhanced by a second showing of the Kew ethnobotanical exhibition specially transported from York. Participants returned to the University of York for the Conference Dinner, featuring traditional Yorkshire fare, which all agreed to be a fitting end to an unforgettable day. The Conference concluded the next day with a further sequence of outstanding lectures.

Conference programme

Monday (20 September)

Morning Session

Chairman: Professor J.G. Hawkes

Romero, G.A. (Cambridge, Mass.) Orchidaceae Spruceanae

Henderson, A. (New York) Richard Spruce and the Palms of the Amazon

Zarucchi, J.L. (St Louis) Contribution of Richard Spruce to our present-day knowledge of the flora of Peru

Afternoon Session

Chairman: Professor M.R.D. Seaward

Crosby, M.R. (St Louis) Richard Spruce's contribution to muscology

Stotler, R.A. (Carbondale, Illinois) Richard Spruce: his fascination with liverworts and its consequences

Gradstein, S.R. (Utrecht) *Hepaticae Amazonicae et Andinae*: an appraisal

Stiff, R. (Wisconsin) Margaret Mee and Richard Spruce

Poster Presentations

Public Lecture, Yorkshire Museum: Prance, G.T. (Kew) A contemporary explorer in the footsteps of Richard Spruce

Buffet Dinner, Yorkshire Museum

Tuesday (21 September)

Morning Session

Visit to Coneysthorpe

Introduction by The Hon. Simon Howard

Remembrance Service, Terrington Church, conducted by the Rector, the Rev. E. Chapman, with contributions by Richard Schultes (Cambridge, Mass.) and Winston Spruce (Wolverhampton)

Tour of Castle Howard (with display of Spruce memorabilia)

Afternoon Session

Lecture Room, Castle Howard

    Schultes, R. (Cambridge, Mass.) Richard Spruce — the man

    Pearson, M. (Doncaster) The early life of Richard Spruce: the making of a naturalist

Conference Dinner, University of York

Wednesday (22 September)

Morning Session

Chairman: Professor R.E. Schultes

    Drew, W.B. (Tubas, Arizona) Spruce's work on *Cinchona* in Ecuador

    Naranjo, P. (Quito, Ecuador) Spruce's great contribution to human health

    Ewan, J. (St Louis) Tracking Richard Spruce's legacy from George Bentham to Edward Whymper [paper presented by M.R. Crosby]

    Madriñan, S. (Cambridge, Mass.) Richard Spruce's pioneering work on tree architecture

    Smith, N.J.H. (Gainsville, Florida) Relevance of Spruce's work to conservation and management of natural resources in Amazonia: perspectives of a geographer

    Reichel-Dolmatoff, G. (Bogotá, Colombia) An anthropologist's debts to Richard Spruce

Afternoon Session

Chairman: Professor G.T. Prance

    Porter, D.M. (Blacksburg, Virginia) Humboldt, Wallace and Spruce at San Carlos de Rio Negro

    Dickenson, J.P. (Liverpool) Bates, Wallace and economic botany in Amazonia, *circa* 1850

    Vreeland, J.M. (Chiclayo, Peru) Richard Spruce in northern Peru: notes on the cultivation of indigenous cotton [paper presented by M.R.D. Seaward].

List of participants (*) and contributors to commemorative volume (‡)

‡ Berry, Paul E. (St Louis, USA)

* Byrn, Richard (Leeds, UK)

* Chadwick, Arthur (Leeds, UK)

* Chávez, Flor (New York, USA)

* Crandall-Stotler, Barbara (Carbondale, USA)

* Crosby, Marshall R. (St Louis, USA)

* Dadd, Michael (York, UK)

* Daly, Juliet (Kew, UK)

* Daly, Michael (Kew, UK)

* Dawson-Brown, Penelope (Pickering, UK)

‡ * Dickenson, John (Liverpool, UK)

* Drew, William (Tubac, USA)

‡ Edwards, Sean R. (Manchester, UK)

‡ * Field, David V. (Kew, UK)

‡ * FitzGerald, Sylvia (Kew, UK)

‡ * Fox, Brian W. (Stockport, UK)

‡ * Gradstein, S. R. (Utrecht, The Netherlands)

* Grant, William (Quebec, Canada)

* Greycloud, Arthur (Tubac, USA)

* Hackney, Paul (Belfast, UK)

‡ * Hackney, Catherine (Belfast, UK)

* Harley, Ray (Kew, UK)

* Hawkes, Jack G. (Birmingham, UK)

‡ * Henderson, Andrew (New York, USA)

* Hennessy, Christopher (Ringwood, UK)

* Hicks, David (Hertford, UK)

* Howard, Simon (Castle Howard, UK)

* Kerr, Graham (Reading, UK)

‡ Lamond, Jennifer (Edinburgh, UK)

* Lauste, Leslie (Brighton, UK)

* Lucas, Gren Ll. (Kew, UK)

* Mackinder, Barbara (Kew, UK)

‡ * Madriñan, Santiago (Cambridge, USA)

* Marsden, John (London, UK)

‡ * Naranjo, Plutarco (Quito, Ecuador)

* Naughton, Gerard (York, UK)

* Naughton, Jane (York, UK)

* Parker, Ken (Hertford, UK)

‡ * Parnell, John A.N. (Dublin, Ireland)

* Paton, Jean A. (Probus, UK)

‡ * Pearson, Michael B. (Doncaster, UK)

* Pickersgill, Barbara (Reading, UK)

‡ * Porter, Duncan M. (Blacksburg, USA)

‡ * Prance, Ghillean T. (Kew, UK)

‡ * Reichel-Dolmatoff, Gerardo (Bogota, Colombia)

* Richards, S. Anne (Cambridge, UK)

‡ * Richards, Paul W. (Cambridge, UK)

‡ * Romero, Gustavo (Cambridge, USA)

* Schultes, Neil (Melrose, USA)

‡ * Schultes, Richard E. (Melrose, USA)

‡ * Seaward, Mark R. D. (Bradford, UK)

‡ * Smith, Nigel J. H. (Washington, USA)

‡ * Spruce, Winston S. (Wolverhampton, UK)

‡ Stiff, Ruth L. A. (Wisconson, USA)

‡ * Stotler, Raymond (Carbondale, USA)

* Thompson, Michael J. A. (York, UK)

* Venturieri, Giorgini A.(Reading, UK)

* Williams, James (Kingston-upon-Thames, UK)

* Willmott, Gary (Hertford, UK)

* Windisch, Paulo (Brazil)

‡ Vreeland, James W. (Chiclayo, Peru)

* Zarucchi, James (St Louis, USA)

Index

Compiled by P.J. Wortley

Note: The limitations of time and space did not permit full indexing of The Bibliography of Richard Spruce, on pp. 303–314, nor full reconciliation of old and modern names of plants and places mentioned in this volume, beyond what is given in the papers presented. Nevertheless, it is hoped that this Index will be a useful guide to the rich diversity contained in this volume on Richard Spruce.